UNIVERSITY OF STRATHCLYDE
30125 00825288 3

D1756476

ANDERSONIAN LIBRARY
WITHDRAWN
FROM
LIBRARY
STOCK
UNIVERSITY OF STRATHCLYDE
UNIVERSITY OF
STRATHCLYDE LIBRARIES

Appraisal and repair of building structures

Reinforced concrete

Acknowledgement

The author is indebted to Dr Roger Browne and Mr Lim Keng Kuok for their assistance during the preparation of this book.

The author also acknowledges K. C. G. Berkeley and S. Pathmanaban and publishers Butterworths for the permission to include the information given in Sections 5.11 and 5.12.

Series Drafting Committee

Other titles in the Appraisal and repair of building structures series:

Appraisal and repair of building structures: introductory guide

Appraisal and repair of claddings and fixings

Appraisal and repair of steel and iron

Appraisal and repair of foundations

Appraisal and repair of timber

Appraisal and repair of masonry

Appraisal and repair of building structures

Reinforced concrete

R. Holland

Published by Thomas Telford Ltd, Thomas Telford House,
1 Heron Quay, London E14 4JD

First published 1997

A catalogue record of this book is available from the British Library

ISBN: 0 7277 2583 1

Throughout this book the personal pronouns 'he', 'his', etc. are used when referring to 'the engineer', 'the client', etc. for reasons of readability. Clearly it is quite possible these hypothetical characters may be female in 'real-life' situations, so readers should consider these pronouns to be grammatically neuter in gender, rather than masculine, in all cases.

Typeset in Great Britain by MHL Typesetting Ltd, Coventry

Printed in Great Britain by The Cromwell Press, Melksham, Wiltshire

Contents

1. Introduction 1
2. Initial appraisal 7
3. Signs of distress 20
4. Principal investigation 40
5. Causes of defects 54
6. Assessing and increasing strength 74
7. Treatment of cracks 81
8. Other repairs to defective concrete 88
9. Repair and replacement of reinforcement 100
10. Protective coatings 104
11. Cathodic protection 113
12. Prestressed concrete 118
13. Floors and roofs 124
14. Basements 134
15. Periodic inspection and maintenance 141
Index 147

1

Introduction

1.1. The myth versus the hard fact

The myth that concrete structures can last forever with little attention needs to be dispelled. Over the years, reinforced concrete has been promoted as having an indefinitely long life requiring negligible attention. Protection for the embedded steel against the external environment was believed to be adequately provided by the thickness of cover and the quality and type of concrete in the cover zone — a view reinforced by the relevant codes of practice. Aggregates and cements were assumed to be mutually compatible.

Although many concrete structures have shown excellent durability for over 50 years, there is now a world-wide problem of deterioration caused primarily by reinforcement corrosion, and expensive diagnostic survey and repair programmes have become commonplace in many parts of the world. Corrosion of reinforcing steel, however, is but one of the many possible problems with which the appraising engineer has to deal.

On the other hand concrete structures, in general, are reasonably tolerant to the inadequacies of designers and builders as well as the misuse and negligence of occupiers. In cases where alternative load paths and high redundancies are present, behaviour in service can be quite different from that indicated by the design calculations and deteriorated structures have been seen to stand up well for several years. Serious disasters arising from distress in structures of relatively young age do however occur.

1.2. The objective and scope of this supplementary guide

The primary objective of the present volume is to provide practical and concise guidance for the practising civil or structural engineer engaged in the appraisal and repair of reinforced concrete building structures. It is not however intended to be a textbook on the subject and is guided by two basic principles. Firstly, it is assumed that the reader has a good structural engineering background and a knowledge of the terms used in building. Secondly, repetition of information published elsewhere is avoided except where such information is necessary for completeness and easy reference. The attention of the reader is however drawn to what the author considers to be the best readily available sources of more detailed information, both in the text and in the select bibliographies which follow each chapter.

This guide is one of a series dealing with the appraisal and repair of building structures in which the Introductory Guide to the series (10 in bibliography) provides general information and guidance (including select bibliographies) which will not normally be repeated in this volume. Furthermore, items which do not form part of the building structure (e.g. suspended ceilings, façades, cladding, curtain walls) are not dealt with in detail.

Precast concrete 'systems'

The author has chosen not to separate the treatment of precast structures from the treatment of reinforced concrete building structures in general. Three factors influenced this decision. The first was that almost everything written about the latter applies to the former; the second that exceptions to this general rule tend to apply only to easily recognised 'systems' of building (e.g. Airey houses, large panel structures); and the third that the excellent publications of the Building Research Establishment cover the specialist needs so admirably. As the reader's attention has now been drawn to the existence of the BRE work, it is not considered necessary to refer to each of their publications on 'systems' precast concrete building structures in the bibliography.

Pulverised fuel ash concrete

This guide does not differentiate between concretes made using different types of cements or using cement replacement materials such as pulverised fuel ash (pfa). Attention is however drawn to a series of publications of the Building Research Establishment relating to the durability of pfa concrete (BRE Reports 294, 295, 296; 4, 5, 6, in bibliography).

European practice

While this guide is written for engineers practising in the United Kingdom, many of the techniques described were first developed elsewhere. Readers therefore may be interested in a publication soon to be released by the European Standards body CEN (available through the British Standards Institution) ENV 1504–9 *Products and systems for the protection and repair of concrete structures — definitions, requirements, quality control, evaluation of conformity. Part 9: General principles for the use of products and systems.*

1.3. Overall process of structural inspection and appraisal

The objectives of a structural inspection and appraisal are obviously dictated by the client's brief but, in almost all cases, there is an underlying requirement for the following questions to be answered.

- What is the present state of deterioration and structural distress?
- Is the structure adequate for its intended purpose in respect of strength, serviceability and durability?
- What will be the future state of deterioration and structural distress?
- What, if any, remedial works are required now or in the future?

In order to answer these questions the engineer is advised to follow a systematic approach of inspection and appraisal such as shown in Fig. 1.1.

Although the approach to the appraisal of a simple structure and that for one of more complexity will differ considerably the

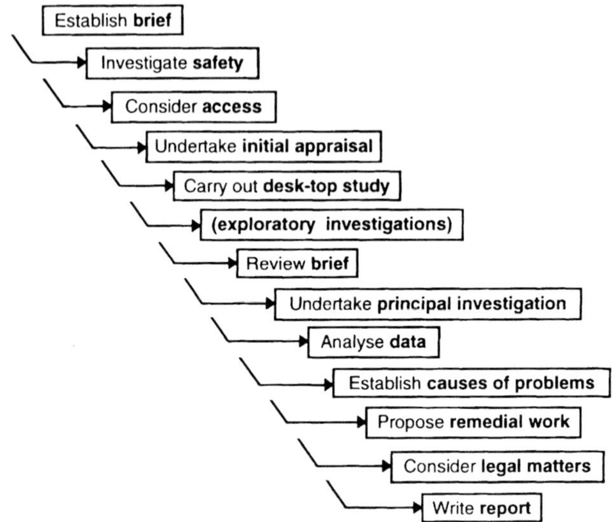

Fig. 1.1. Appraisal procedure

following headings, based on Fig. 1.1, will provide a comprehensive checklist or outline procedure.

- The brief
- Safety
- Access

Guidance on all three of these subjects is given in the Introductory Guide to the series (10 in bibliography).

- Initial appraisal — guidance is given in Chapter 2 of this volume and Chapter 3 includes advice on signs of structural distress.
- Desk-top study — time should then be taken to consider the information obtained so far, to review the brief and to plan the remainder of the appraisal in detail. Although the initial appraisal may have indicated the general direction in which investigations should proceed, it is nevertheless prudent not to discard alternatives until all reasonable objections to them have been validated. This may require further background information and it may be necessary to consult others with particular experience in appropriate fields. The Introductory

Guide to this series gives advice regarding historic buildings and unusual materials or structures.

- Exploratory investigations — before a programme of extensive investigations or tests are embarked upon it is normally prudent to carry out a series of exploratory tests. The objects of these would be to highlight access and other difficulties which may not have been foreseen initially, to confirm any previous cost and time estimates and to establish that the tests envisaged are the most appropriate.
- Review of the brief — if the information gathered so far indicates that changes to the brief will be required, the client should be contacted for his instructions.
- Principal investigations — the principal investigations can then be put in hand. Guidance on methods of investigation is given in Chapter 4 of this volume and on instrumentation and monitoring and the load testing of structures in the Introductory Guide to the series.
- Analysis of the data — The Introductory Guide (10 in bibliography) to the series gives guidance on assessing the strength of existing structures and Chapter 6 of this volume addresses issues especially relevant to reinforced concrete.
- Establishing the causes of failure/deterioration — guidance on causes of failure can be found in Chapter 5 of this volume.
- Remedial measures/maintenance required — Chapters 7 to 14 of this guide give advice on remedial measures and Chapter 15 on maintenance.
- Legal and other non-engineering matters — The Introductory Guide to the series gives advice on legal aspects and deals with the special requirements for historic buildings.
- Report — The Introductory Guide to the series gives advice on writing the report.

Select bibliography

1. Allen R. T. L. *et al. The repair of concrete structures*. Blackie, London, 1993, 2nd edn.

2. American Concrete Institute. *Guide for making a condition survey of concrete in service*. The Institute, 1993.
3. American Concrete Institute. *Manual of concrete inspection*. The Institute, 1992, ACI Special Publication 2, 8th edn.
4. Building Research Establishment. *Performance of pfa concrete in aggressive conditions: 1, Sulphate resistance*. BRE, Garston, 1996, BRE Report 294.
5. Building Research Establishment. *Performance of pfa concrete in aggressive conditions: 2, Marine conditions*. BRE, Garston, 1996, BRE Report 295.
6. Building Research Establishment. *Performance of pfa concrete in aggressive conditions: 3, Acidic groundwaters*. BRE, Garston, 1996, BRE Report 296.
7. Building Research Establishment. *Repair and maintenance of reinforced concrete*. BRE, Garston, 1994, BRE Report 254.
8. Campbell-Allen D. and Roper H. *Concrete structures: maintenance and repair*. Longman, London, 1991.
9. Comité Euro-International du Béton . *Durable concrete structures — design guide*. Thomas Telford, London, 1992.
10. Holland R. *et al.* (eds). *Appraisal and repair of building structures — introductory guide*. Thomas Telford, London, 1992.
11. Kay T. *Assessment and renovation of concrete structures*. Longman, London, 1992.
12. Mays G. (ed.). *Durability of concrete structures — investigation, repair, protection*. Spon, London, 1992.
13. Perkins P. H. *Repair, protection and waterproofing of concrete structures*. Elsevier. London, 1986.
14. Puller-Strecker P. *Corrosion damaged concrete — assessment and repair*. Butterworth, London, 1987.

2

Initial appraisal

2.1. Introduction

The purpose of the initial appraisal is to gain a feel for the nature and extent of the work to be carried out and to assist in agreeing the final brief with the client.

The objectives should be

- to prepare an agreed brief for the full appraisal
- to gain an understanding of the main problems likely to be encountered
- to obtain sufficient information to forecast the resources needed for the full appraisal
- to plan any specialist services.

An initial appraisal, sometimes referred to as a global survey, can be initiated as a result of complaints by owners, occupiers or members of the public who are alarmed at cracks or spalling which they recognize as signs that something has gone seriously wrong with a reinforced concrete structure. It may also result from the findings of a periodic inspection of the building structure. Of course, a building with no apparent defects may be similarly subjected to such scrutiny if the client, for some reason, so requires — for example, when there is a proposed change of use or additions or alterations to the building which are likely to result in an increase in loading. An initial appraisal is the first step

Signs of distress!

towards establishing the feasibility of such proposed changes.

The initial appraisal aims to provide information about the structure as a whole and normally consists of a visual condition survey combined with limited non-destructive testing and core sampling of the concrete where found to be necessary. A close-up visual inspection should, as far as possible, be carried out over the entire surface of the structure. It should seek to confirm the existence of any defects of significance with regard to the durability and structural integrity of the building. It should also lead to the identification of other areas of concern and result in recommendations for any more detailed survey or comprehensive investigation which the engineer considers to be necessary or advisable. Immediate actions should, of course, be recommended for areas where distress is serious or danger imminent.

2.2. *Preparations for the initial appraisal*

Before carrying out an initial appraisal it is crucial to plan the operation and gather together all available information relevant to the structure. This information, which may be thought to be of little significance initially, may hold vital clues to the performance of the structure and may include

- original structural calculations, detailed as-built drawings and photographs taken during construction
- test records for aggregates, cement and concrete, reinforcement and other materials
- historic records of the use of the structure and also of the site prior to construction
- climatic and environmental records and details of atmospheric or ground contamination.

In addition it is essential to establish the detailed requirements of the owner, his intended future use for the structure and its required residual life, which may prove to be quite different from the original design intention.

The next stage in planning the appraisal is the development of a field specification. This is a planning document which advises both the client and the members of the engineer's team of the survey objectives, methods and scope. It should contain all the information pertinent to the survey including previous survey results, maintenance and repair records. It should also contain proforma sheets which can be used to record the mass of data in a logical manner. These consist of a plan of each element to be surveyed, a drawing to locate the element within the overall structure, and a reference system to relate all the test data to the element.

Generally, the engineer will only have limited time on the site to conduct this survey so it is essential that he is familiar with the layout and general structural details as shown on the drawings, design calculations, records of construction and the history of maintenance. These documents may also indicate areas warranting special attention and assist in locating the limited initial tests required and the type of equipment to take to the site. This may be particularly important in the case of a distant site where repeat visits might be costly, time-consuming or not feasible.

2.3. *Tools for the initial appraisal*

Planning what tools and equipment will be required for an initial appraisal is an important prerequisite for a successful survey. Spares may be necessary for some of the more important tools.

The following are normally required in the initial appraisal inspection of a reinforced concrete structure

General

- field specification (see above)
- selected drawings
- sketches and proformas for recording data
- notebook and pencils
- appropriate and safe apparel.

Visual conditions survey

- camera (SLR with wide angle and telephoto lenses)
- photographic flash and films
- binoculars
- powerful torch
- measuring tape and steel ruler
- spirit level
- crack width measuring device
- boroscope.

In-situ tests

- hammer for tapping surfaces
- cover meter
- bottle of phenolphthalein and brush
- electric drill, extension cable and adaptors
- cold chisel.

2.4. *Identifying the areas of concern*

A prime task during the initial appraisal is to identify, for more comprehensive investigation, critical regions of the structure which are particularly vulnerable to structural instability or deterioration.

These may include

- areas of low redundancy where local weakness due to deterioration or strength deficiency could result in instability of the whole structure
- areas of inadequate robustness where deterioration or accidental loads could result in progressive collapse of the structure
- areas subjected to high stress either during construction or in service
- areas of potential weakness resulting from difficulties during construction

Exposure — the decorative crown of this building deteriorated significantly faster than the main structure

- areas subject to high environmental loading (e.g. wind, snow) or particularly aggressive environments
- areas where deterioration is already extensive.

Areas where potential defects or weaknesses can reasonably be suspected should also be included — for example

- where rust staining or spalling are visible and have been caused by insufficient concrete cover it is likely that adjacent areas which do not, as yet, show such visible signs may have related problems
- if structural cracks are observed in some components it is likely that similar components subjected to comparable loading are potential candidates for further investigation.

2.5. *The visual conditions survey*

The visual conditions survey is perhaps the most important part of any structural investigation. It aims to provide information on types, extent and seriousness of visual phenomena, both on the structure and on other parts of the building, which may result from defects in the structure. The results will be found particularly valuable in determining whether further investigations are necessary in order to identify the causes of defects or to assess their extent.

Sound engineering judgement, experience and an understanding of structural forms, load paths, and the significance of unusually long spans, unusual member sizes and proportions, critical connections etc. are the essential characteristics required of an inspection team. It follows therefore that the work should be entrusted only to experienced civil or structural engineers, or to experienced personnel directly under their control. An experienced engineer, approaching the survey with an open mind, is best qualified to assess the condition of the structure objectively. A mere visual effort, without any structural understanding or consideration, will lack the necessary depth on which to make further decisions and may result in misleading conclusions.

A common dilemma facing an engineer during a visual condition survey is the extent to which structural elements — and hence, possibly their defects — are hidden by architectural finishes both within and on the exterior of a building. Soffits of concrete beams and slabs are often above ceilings; columns and walls behind tiles, marble or timber panelling; floor surfaces under tiles, carpet or other floor finishes; external members covered by cladding or curtain walls. Such constraints and limitations may be recognized when carrying out a survey of a domestic property for valuation but, if a full picture of the structural health of a building is required, an inspection of the hidden structure, to a certain extent, must be carried out but not without first obtaining the specific consent of the owner. Direct inspection of structural elements at selective locations can often be achieved by opening up ceiling panels or by the removal of small areas of architectural finishes from locations which are not prominent. The extent and locations of these selective inspections will depend on engineering judgement, experience and the particular circumstances of the case. A preview of the design, construction records, as-built drawings, maintenance records and observations during inspection would help decide on where and how much to inspect.

Generally, direct inspection should be considered for areas where

- signs of distress are evident in the architectural finishes or in adjacent areas of the structure
- construction records have drawn attention to errors in construction or to deficiencies in the quality of works or materials
- maintenance records have indicated previous repairs, strengthening or reports of signs of distress
- structures are critically exposed to agents of degradation
- structures are critical because of long spans, small member sizes or heavy loads
- significant areas of the structure would otherwise remain un-inspected.

The engineer should also be aware that the structural integrity of the building may be affected in the following situations

- when areas do not appear to be part of the original building or appear to be out of place — many additions or alteration

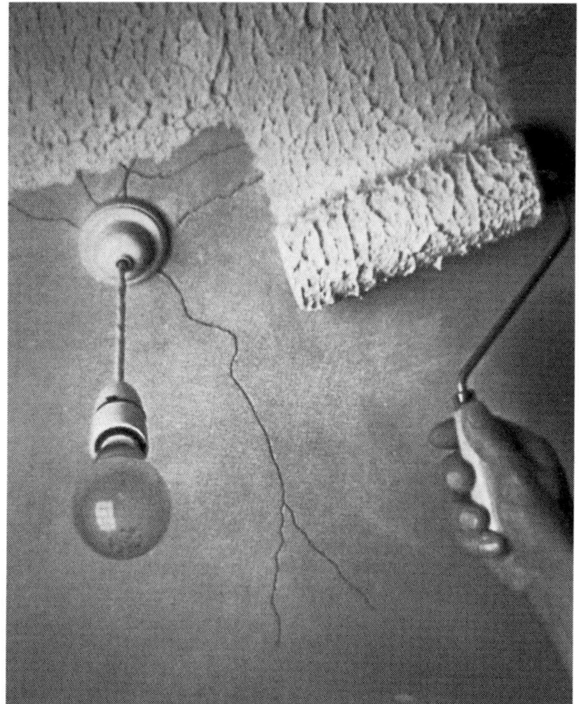

Newly decorated surfaces may hide evidence

works can be detected in this manner (e.g. an additional mezzanine or attic floor constructed of a different material; a new roof converting the original reinforced concrete roof slab to a floor; new and possibly heavier roof coverings)

- when loadings appear to be abnormal or excessive (e.g. a unit for goods or document storage in a residential block; unusually heavy machines in a light factory building)
- when previous strengthening or repairs have been carried out (e.g. a brick pier or steel prop may provide additional support to a deficient beam; an unusually large column may have been modified to increase the capacity of a deficient column). Sometimes, new or different architectural finishes will give clues to modification works done (e.g. a patch of new tiles around a column base may well be evidence of underpinning work or the enlargement of an existing foundation).

2.6. *Making records*

An essential adjunct to the visual conditions survey is the systematic recording of the observations made. In order to speed up the survey process and to achieve consistency between different members of the inspection team regarding the interpretation of visual defects, a classification system can be used. Recording should be done on proformas which are prepared before the survey (see Section 2.2 above). Where appropriate, a computer database can be developed to maintain the records of the observations.

A comprehensive photographic record of the condition of the structure at the time of inspection, with cross reference to the sketches and text of the report, is important and serves as a base for comparisons with future surveys. For this reason, photographs should be taken both of areas with and without defects, particularly where the former are of critical structural importance. The use of video equipment to make a simultaneous visual and audio report on the structure may also be of value.

The value of such records as a contribution to the overall management and maintenance of the structure during its life can be significant.

2.7. *Non-destructive tests and core samples*

A visual inspection will enable only detection of defects, deterioration or structural weaknesses that are sufficiently severe to be apparent at the concrete surface at the time of the inspection. Limited non-destructive tests, such as those listed in Table 2.1, aimed at identifying factors such as concrete strength, internal voids or corrosion, should therefore be considered at selected locations as part of the initial appraisal.

The types, number and locations of these non-destructive tests would depend on the nature and seriousness of defects observed and on whether a repeat visit would be prohibitively costly or time-consuming.

Where durability problems are suspected it is useful to detect delamination by hammer tapping. Some tests on carbonation depth, chloride content, thickness of concrete cover and quality of the cover concrete can be performed on selected areas to assess the severity of the problem and to decide whether further tests are necessary.

Table 2.1 Common non-destructive tests used during initial appraisal

Properties examined	Non-destructive testing techniques	Other considerations
Carbonation depth	Phenolphthalein	Concrete broken out to give fresh exposed surface (also cores used)
Chloride content of concrete	'Quantab' test strip	Site test on powdered sample (6 in bibliography)
Thickness of concrete cover	Covermeter	
Quality of cover concrete	Impact tests (e.g. Schmidt hammer)	Indicates quality close to surface
	Ultrasonic pulse velocity (e.g. Pundit)	Locates cracks and indicates quality throughout cover zone using variable spacing of heads
Lamination from corrosion of reinforcement	Hammer tapping	Clearly indicates corrosion expansion cracks
	Endoscope	Size and depth of cracks
Concrete strength	Ultrasonic pulse velocity (e.g. Pundit)	Direct or indirect
	Impact tests (e.g. Schmidt hammer)	

For the purpose of estimating in-situ concrete strength, a comprehensive non-destructive testing scan of the structure, to identify those areas with material weaknesses and the range of in-situ strength, can overcome the impracticability of taking many core samples at this stage. This test scan will also give a fairly good idea of the variability of the quality of the concrete in the structure.

Laboratory tests on limited numbers of small core samples of 50 mm diameter can give a preliminary idea of the quality of the concrete in terms of strength, density, mix proportions and chemical contamination.

2.8. *Interpreting the findings*

As defects often arise from a combination of causes it is important to keep an open mind in the interpretation of the findings at this stage of the investigation.

The observed condition of the concrete may give a false picture as regards the condition of the structure as a whole; for example, a badly deteriorated reinforced concrete structure may not necessarily mean an unsafe one. Further analysis may show the deteriorated structure to be satisfactory from the strength point of view but require immediate repairs to enhance serviceability and durability. On the other hand a structure with no obvious signs of deterioration may be suffering from a lack of adequacy in stability or robustness.

Interpretation of the observations and of the results gained from the limited tests carried out during an initial appraisal depends upon the experience, expertise and judgement of the engineer who is responsible for conducting the investigation. It may also require the advice of experts in other fields (e.g. materials science, concrete and corrosion technology) in the interpretation of the test results. One of the best published sources of critical guidance on the use and interpretation of the many tests available is given in J. H. Bungay's *The testing of concrete in structures* (7 in bibliography).

2.9. *Conclusions and recommendations*

On completion of the initial appraisal the engineer should be in the position to

- recommend immediate actions (e.g. propping, shoring, evacuation) if the structure is found to be in imminent danger of collapse
- define areas of the structure for further investigation

- specify the most appropriate testing techniques to be used in the principal investigation
- postulate the causes of defects, whilst keeping an open mind until completion of further investigations
- estimate the cost of the principal investigation.

Where the initial appraisal finds no defects of immediate significance affecting durability or strength the engineer will be in a position to submit a report recommending any minor repairs or monitoring that may be considered necessary. The importance of periodic inspection and maintenance (see Chapter 15) should be emphasized.

Select bibliography

1. Allen R. T. L. *et al. The repair of concrete structures.* Blackie, London, 1993, 2nd edn.
2. American Concrete Institute. *Guide for making a condition survey of concrete in service.* The Institute, 1993.
3. American Concrete Institute. *Manual of concrete inspection.* The Institute, 1992, ACI Special Publication 2, 8th edn.
4. British Standards Institution. *Testing concrete,* BS 1881, Part 201: *guide to the use of non-destructive testing for hardened concrete.* BSI, London, 1986.
5. British Standards Institution. *Testing concrete,* BS 1881, Part 202: *surface hardness testing by rebound hammer.* BSI, London, 1986.
6. Building Research Establishment. *Simplified method for the detection and determination of chloride content in hardened Portland Cement concrete.* BRE, Garston, 1977, BRE Information Paper IP 12/77.
7. Bungay J. H. *Testing of concrete in structures.* Blackie, London, 1989, 2nd edn.
8. Concrete Society. *Non-structural cracks in concrete.* The Society, London, 1982, Technical Report 22.
9. Fédération Internationale de Précontrainte (FIP). *Inspection and maintenance of reinforced and prestressed concrete.* Thomas Telford, London, 1986.
10. Holland R., *et al.*, (eds). *Appraisal and repair of building structures — introductory guide.* Thomas Telford, London, 1992.
11. Institution of Structural Engineers. *Appraisal of existing structures.* The Institution, London, 1996.

12. Institution of Structural Engineers. *Guide to surveys and inspections of buildings and similar structures*. The Institution, London, 1991.
13. Mays G. (ed.). *Durability of concrete structures — investigation, repair, protection*. Spon, London, 1992.
14. Perkins P. H. *Repair, protection and waterproofing of concrete structures*. Elsevier, London, 1986.
15. Puller-Strecker P. *Corrosion damaged concrete — assessment and repair*. Butterworth, London, 1987.

3

Signs of distress

3.1. Looking for signs of distress

There is often a tendency to equate signs of distress in concrete with cracking. It is however important, while carrying out an investigation, to recognize that evidence of distress in reinforced concrete structures is not limited to cracks. Other signs such as excessive deflections in beams and slabs or bowing of columns are equally important. Rust stains, surface deposits, dampness or leakages are further indicators of possible problems. While some of these signs are of structural (strength) concern, others are related to durability (serviceability). If neglected, defects relating to serviceability could deteriorate to a stage when structural integrity becomes affected.

The structural appraisal should not be limited to the observation of concrete elements alone. The first signs of structural deficiency are often reflected in non-concrete components due to interaction between the different parts of a building — for example, cracks in brickwork or blockwork walls, buckling of partitions, cracking of glass in windows, distortion of door frames. It is also important to consider those cases where concrete members are concealed by architectural finishes and where access for inspection is limited or non-existent. Removal of finishes may not always be possible but careful scrutiny of the architectural finishes and adjacent exposed concrete elements would, in many cases, enable adequate judgement and conclusions to be made on the condition of the structure.

3.2. *Categorization of common defects and signs of distress*

Defects and signs of distress in reinforced concrete structures commonly fall under the following broad categories

- on Concrete Elements
    - cracking
    - surface deterioration
    - surface deposits
    - deformation
    - construction defects
    - construction features.
- on Non-Concrete Components
    - cracking and distortion.

In the remainder of this chapter, using the sequence outlined above, the appearance and nature of defects and their associated signs of distress are discussed. It should be remembered, however, that the manifestation of the observed signs can often be the result of a combination of various causes happening consecutively or simultaneously. The reader should note that the cause of distress only will be considered in this chapter as an aid to detection: detailed discussion of the causes of defects will be found in Chapter 5.

3.3. *Cracking in concrete*

The diagnosis of cracks is a complicated matter requiring careful site observations, good understanding of the mechanisms leading to the formation of cracks, intelligent interpretation of test results, experience and engineering judgement. The subject is considered of sufficient significance to insurers that the Loss Prevention Council has issued guidance which the investigating engineer is advised to take into account (22 in bibliography).

Concrete has a relatively low tensile strain capacity which is a function of the strength of the concrete, its age and the rate of application of the strain. Cracking will occur whenever the tensile strain to which the concrete is subjected exceeds its tensile strain capacity.

It is to be emphasized that observed cracks can sometimes be the resultant effect of various causes during the life of a structure and therefore have to be interpreted with care. For instance, cracks originating from plastic settlement during the construction stage might have gone un-noticed until corrosion of reinforcement takes place years later. Dirt in a crack may indicate that it is not of recent origin.

The pattern which cracks form and their location are probably the two most important clues to their cause. Straight cracks may indicate some association with reinforcement whereas a mesh type of pattern could point towards drying shrinkage, surface crazing, frost attack or alkali–silica reaction. The products of cracks can also help to narrow down the possible cause, as in the case of an exuding gel associated with alkali–silica reaction and the minerals or salts deposited by water passing through cracks or porous concrete. Table 3.1 and Fig. 3.1 are provided as a quick guide to the probable causes of observed cracking but should not be used in isolation from the text of this guide. In the table the term 'shift' is used to indicate permanent lateral displacement across the crack and 'movement' to indicate that the crack opens and closes in response to some external stimulus. The table may also serve as a check list indicating what information to record in relation to cracks.

The significance of cracks

It must be remembered that the fundamental principle of reinforced concrete requires cracking to occur within certain limits for the reinforcement to take its effect in tension. The widths of such cracks can be limited to acceptable levels by appropriate detailing in accordance with codes of practice. BS 8110 (4 in bibliography), for example, allows crack widths of up to 0.3 mm. BS 8007 (3 in bibliography) limits crack widths for direct tension and flexure in water retaining structures to 0.2 mm for severe or very severe exposure and 0.1 mm where aesthetic appearance is critical.

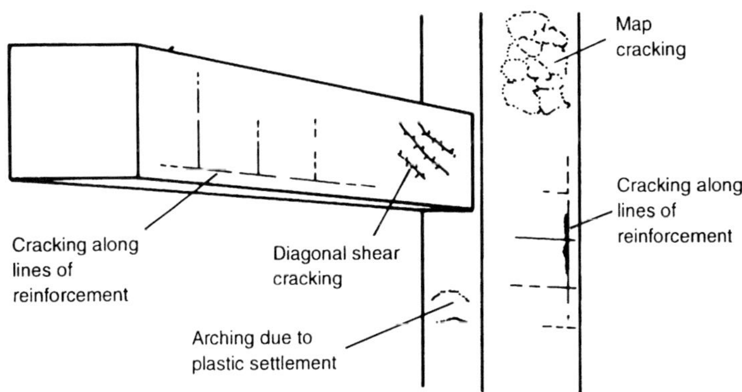

Fig. 3.1. Common crack patterns

Concrete is also liable to cracks in both the plastic and hardened states owing to the internal tensile stresses that arise from the response of its constituent materials to environmental effects and restraints to these effects. In the majority of cases, such non-structural or intrinsic cracks are quite harmless to the structure as a whole, except where they provide access to the reinforcement through which corrosion might be initiated.

Leakage and 'autogenous healing'

Research has shown that, under certain favourable conditions, cracks less than 0.2 mm wide are capable of a process called 'autogenous healing'. During this process, bonding materials are formed and deposited in the cracks, sealing them and preventing further leakage. Healing will not be possible if a crack is disrupted by movement or a flow of water occurs which is sufficiently great to wash out the deposits.

Cracks and the corrosion of reinforcement

When cracks occur as a result of expansion of the corrosion products of steel reinforcement the line of the crack will, in almost every case, follow the lines of the corroded reinforcement. On the other hand, when cracks are due to other causes, it is often difficult to predict what effect they are likely to have on the corrosion of

Table 3.1 Cracking in concrete — diagnostic chart

Pattern	Location	Width	Depth	Shift	Movement	Environment	Products	Other	Probable cause
Arching	Narrow vertical members	0–10 mm	Deep	None	None	Any	None		Plastic settlement
Diagonal	Slabs	2–3 mm	Tapering to zero	None	None	Any	None	Single or parallel	Plastic shrinkage
Diagonal	Ends of beams	Hair-line	Through section	Possible	Unlikely	Any	None		Shear cracks
Diagonal	Around member	Varies	Perimter area only	None	Unlikely	Any	None		Torsion cracks
Diagonal	Walls, beams	Varies	Through section	Possible	Progressive	Any	None	Subsidence	Ground movement
Irregular	Floor slabs	Varies	Through section	Likely	Progressive	Any	None	Subsidence	Ground movement
Irregular	Edge of structure	Varies	Through section	Likely	None	Any	None	Impact scars	Impact
Irregular	Foundations	Progressive	Progressive	None	None	Damp	Whitish appearance		Sulphates attack
Irregular	Mosaics, brick slips	Small	Through finishes	Possible in finishes	None	Any	None		Adhesives failure
Long	Over reinforcement	0–5 mm	To reinforcement	None	None	Any	None	Taper	Plastic settlement
Long	At change of geometry	0–10 mm	Deep	None	None	Any	None		Plastic settlement
Long	Any	0–10 mm	Through section	Unlikely	Seasonal	Any	None		Thermal movement

Long	Slabs	2–3 mm	Deep	None	None	Any	None		Plastic shrinkage
Long	Thick walls or slabs	3–7 mm	Any	None	None	Any	None	Ceases at joints	Early thermal contraction
Long	Any	Progressive	To reinforcement	None	None	Humid or damp	Rust in later stages		Reinforcement corrosion
Long	Beams, slabs	Varies	Tension zone only	Unlikely	With load change	Any	None	Across line of force	Flexural cracks
Long	Rare	Varies	Through section	None	With load change	Any	None	Across line of force	Tension cracks
Map (mesh) -like	Any	0–5 mm	Shallow	None	None	Any	None		Surface crazing
Map (mesh) -like	Thin slabs or walls	2–3 mm	Deep	None	None	Any	None		Drying shrinkage (long term)
Map (mesh) -like	Any	Progressive	Progressive	None	None	Damp	Gel		Alkali–silica reaction
Map (mesh) -like	Any	Varies	Progressive	None	None	Exposed	None	On poor concrete	Frost damage
Parallel	Members in compression	Varies	In compression zone	None	Unlikely	Any	None	Across line of force	Compression cracks
Random	Slabs	2–3 mm	Deep	None	None	Any	None		Plastic shrinkage
Short	With flexural cracks	Varies	To reinforcement	None	None	Any	None		Bond failure
Vertical	Masonry, tiles	Varies	Through finishes	None	None	Any	None		Creep

reinforcement. There is also, as yet, no reliable established relationship between crack widths and the onset of corrosion. On the other hand, there is little evidence that cracks less than 0.3 mm wide, whether parallel with or transverse to the reinforcement, can pose any great risk of corrosion although they may, if running along the line of a reinforcing bar, be the first sign that corrosion is already taking place.

In general, longitudinal cracks running along the lines of reinforcement are more likely to lead to corrosion than transverse cracks as the latter, which tend to be the result of flexure, usually taper from maximum at the concrete surface to near zero width at the reinforcement. Before dismissing transverse cracks as a potential threat to reinforcement, however, consideration must be given to the exposure of the structure and the locations and widths of the cracks.

Cracks and structural capacity

Cracks which are caused by overloading or deficiencies in the design, detailing or construction are likely to pass completely through the structure and the width of the cracks will be greater than the limits imposed by the codes of practice and, in some cases, there will be periodic movement along or across them. Their locations and directions are indicative of the nature of the structural deficiencies which caused them, as also is any lateral shift of the concrete surfaces across a crack.

3.4. Volume change related cracks

Volume change in concrete occurs as a result of temperature variation or shrinkage of the concrete. It is a natural process starting minutes after the placing of the concrete and extending for years into the service life of the structure. When restrained, any volume change causes tensile strains often of sufficient magnitude to produce cracks. The types of restraint that can lead to cracking include embedded reinforcement and the forces caused by differential temperature gradient within the concrete mass. Cracks caused by volume change are also referred to as intrinsic cracks.

Plastic settlement cracks

Concrete which is not densely placed or from which there is excessive bleeding of cement paste through formwork or construction joints may settle and so allow voids or cracks to form. Common locations for such cracks are behind permanent formwork — particularly wood wool slab and other porous insulation panels — and at the corners of columns. Plastic settlement cracks, which occur within three hours of concrete being placed, can result in serious corrosion unless treated soon after their formation.

They are of three types.

- Over reinforcement — cracks along lines of and above the top reinforcement in *deep sections*, which usually taper to zero at the ends.
- Arching — cracks across *narrow columns or walls* caused by concrete arching between forms and reinforcement.
- Change of depth — cracks which occur where there is a *sudden change in the geometry of the concrete section* in, for example, trough and waffle slabs.

Plastic shrinkage cracks

Plastic shrinkage cracks in concrete slabs occur within six hours of the concrete being placed and can be distinguished from long-term drying shrinkage cracks not only by their time of appearance but because they rarely reach the edge of the slab. Unlike plastic settlement cracks they usually pass through the whole section. Although plastic shrinkage cracks are, in most cases, structurally harmless they can cause corrosion if the reinforcement becomes exposed to salt or contaminated dust.

Plastic shrinkage cracks are of three types.

- Diagonal — a single crack, or several parallel cracks, of from 2 mm to 3 mm in width at the surface tapering towards the concrete core running *diagonally across concrete slabs*.
- Random — several cracks in a *reinforced concrete slab* showing no regular pattern.
- Over reinforcement — cracks along the lines of reinforcement

in a *reinforced concrete slab* and usually passing through the full depth of the section.

Early thermal contraction

Cracks of this kind occur only in thick sections and rarely pass completely through the whole thickness of the concrete. There is less likelihood of early thermal cracking in concrete where a substantial proportion of the cement has been replaced by pozzolanic materials.

Early thermal cracks are of two types.

- Caused by external constraint — cracks in *thick walls* which have been cast onto an already hardened base, or against an adjacent wall panel without movement joints — the crack will terminate at a construction joint.
- Caused by internal constraint — cracks in *thick slabs* where the interior of the slab has cooled at a slower rate than the surface.

Long term drying shrinkage

Tension cracks may appear in *thin concrete slabs or walls* several

Early thermal contraction

weeks or months after placing of the concrete due to excess water in the mix or inadequate provision of movement joints.

Surface crazing

Surface crazing results from the differential shrinkage of the concrete surface relative to the internal core during or after hardening. It does not result from reinforcement corrosion (see also Sections 3.6 and 3.8 below).

Thermal movement

Changes in temperature can create expansion and contraction of concrete and other materials. If the structure is unable to safely accommodate the movements or strains caused, cracking may result. (Methods of calculating thermal movement are given in 7 and 26 in the bibliography.)

3.5. Corrosion related cracks

Corrosion of steel is an expansive process and the forces exerted by the expansion of the corrosion products (rust) will eventually disrupt the concrete cover. The first visible sign of reinforcement corrosion is usually a hair-line crack (sometimes accompanied by rust stains) on the concrete surface running parallel and directly in line with the reinforcement closest to the surface. In slabs the cracks will be long and follow the lines of the main reinforcement, whereas in beams and columns it is likely that the first cracks will be over the links. The cracks indicate that the expanding rust has grown enough to split the concrete cover. Indeed, only a small amount of rusting is needed to cause cracks.

There are three principal causes of reinforcement corrosion.

- Carbonation of the cover concrete — a phenolphthalein test (see Section 4.13 below) will indicate the depth to which the cover has carbonated.
- Chloride contamination — calcium chloride was, some years ago, regularly used to speed up the set of concrete — *particularly pre-cast concrete*; a further source of chlorides could be road salt or substances used in the manufacturing processes

within a building. Chemical tests (see Section 4.14 below) will reveal the presence of chlorides.

- Intrinsic cracks (see Section 3.4 above) and deterioration in service (see Sections 3.7 and 3.8 below).

3.6. Chemical reaction related cracks

Cracks can be caused by the tensile strain resulting from expansive forces exerted by the products of chemical reactions within the concrete as, for example, with alkali–silica reaction and sulphates attack, or by weakening of the concrete matrix brought about by chemical change as happens in some High Alumina Cement concretes.

Alkali–silica reaction

Alkali–silica reaction (ASR) cracks radiate from expansion sources and will eventually link up to form a 'map' pattern similar to surface crazing. There is often clear or stained gel exuding from the cracks. ASR, however, only occurs in *damp locations* and requires a particular combination of chert aggregate and a cement having a high alkali content. The amount of reinforcement and the magnitude of externally applied stress may change the crack pattern. For example, cracks due to ASR in axially loaded columns are usually longitudinal. Spalling may occur at an advanced stage. BRE Information Paper 16/93 (13 in bibliography) may be of help.

Sulphates attack

As the source of sulphates is normally in the ground it follows that sulphates attack is restricted to elements in contact with the ground such as *foundations*. Cracks due to sulphates attack usually starts at the edges and corners, followed by irregular cracking and spalling of the concrete. This allows further penetration and disruption to the concrete. The concrete has a characteristic whitish appearance and cracks and spalling become deeper with time and lead eventually to complete disintegration. Further information is given in BRE Report 164 and Digest 363 (11 and 12 in bibliography).

High alumina cement

High alumina cement (HAC) is a useful refractory material and until about 1970 it was also used in building structures although almost exclusively for precast elements such as beams and floor units. Unfortunately the grade of HAC then in use, under certain conditions, changes its chemical structure with time which reduces the compressive strength of the concrete. Signs of distress are the same as for strength related defects (see below) and a chemical test would be necessary to differentiate from other types of cement product. Further guidance is given in BRE Digest 392 (15 in bibliography).

3.7. *Strength related cracks*

Cracks can result from excessive stress levels in the concrete. Overloading beyond the design provision, poor construction or deterioration due to environmental factors can result in in-situ strength deficiencies in concrete, reinforcement and their interfaces. In addition to the materials, the structural forms themselves can be lacking in strength and robustness due to inadequate design, detailing or construction. Cracks may also result from stresses due to imposed deformations such as differential settlement of foundations or creep.

It should not be overlooked that an apparent strength deficiency may be due to the structure behaving in a way not anticipated by the designer as, for example, when distress is caused by movement joints failing to operate effectively.

Such deficiencies include

- Deficiencies in shear strength — Diagonally inclined hair-line cracks at or near the supports of beams may indicate insufficient shear strength.
- Deficiencies in flexural strength — Although excessive deflection is the principal indicator of inadequate flexural strength, cracks in the concrete tension zone running transversely to the line of force should not be disregarded in this respect unless they are clearly due to some other cause.
- Deficiencies in tensile strength — Reinforced concrete is not normally designed to carry direct tension but some change in

loading patterns or damage to the structure may cause tensile stresses to occur. Any associated cracks will run transversely to the line of force.

- Deficiencies in compressive strength — Cracks in line with a concentrated compressive load may indicate a deficiency in compressive strength.
- Deficiencies in torsional strength — Diagonal cracks around the perimeter of a section may indicate deficiency in torsional strength.
- Deficiencies in bond strength — Flexural stress cracks (see above) associated with short cracks along the lines of reinforcement may indicate bond failure.
- Overload — Careless overloading can occur during construction or use but the direct evidence may have been removed. Accidental overloading can occur due to impact, explosion, earthquake or wind. Serious overloading can be caused by change of use or ill thought through modifications to the structure.
- Differential settlement — Foundation or soil inadequacy, uneven pressure on the ground due to loads from the structure or frost heave, may impose forces on the reinforced concrete to which it can only respond by cracking.
- Creep — Concrete which is loaded compressively will shrink but most buildings are too small for this to be a significant cause of distress. In tall buildings however the amount of creep can be substantial and, if not allowed for in design, could lead to disruption of the concrete particularly where parts of the structure of differing height are rigidly joined together. (Methods of calculating creep are given in 7, 19 and 26 in the bibliography.)

3.8. Surface deterioration

The following forms of deterioration may occur:

- Pop-outs — Pop-outs are small defects consisting of shallow, conical depressions on concrete surfaces. They are usually caused by the expansion of highly absorptive materials in coarse aggregates which are susceptible to frost action. Sometimes contaminants consisting of clay particles or lignite in the aggregates can also expand to cause pop-outs.

- Spalls — Spalls are fragments of concrete which detach from a surface due to exertion of internal pressure developed under the surface area. This pressure is commonly the result of expansion caused by the corrosion of steel reinforcement. Where gel is evident on the concrete surface the spall is more likely to be the result of the expansive effect of alkali–silica reaction.
- Delamination — Delamination occurs when internal pressure is exerted over a large area causing a sheet spall. In its initial stage delamination may not be immediately apparent on the surface, but simple hammer tapping is very effective in identifying underlying hollowness. In its advanced stage, bulging of the concrete surface is apparent, which is often accompanied by cracks around the delaminated areas. If no action is taken the delamination, if on the face of a wall or column or the underside of a slab or beam, will eventually drop under its own weight.
- Weathering of the concrete surface — Deterioration of the outer skin of a concrete surface can result from the effects of freezing or atmospheric pollution. The extent of the damage which may arise from freezing varies from surface scaling to complete disintegration as layers of ice are formed, first at the exposed surface of the concrete and then progressively through its depth. In the United Kingdom dense well compacted concrete is rarely affected by frost, although reinforced concrete could be damaged by chemicals used to alleviate the effects of frost, the most obvious of which is salt.

  In industrial areas where the atmosphere is polluted with sulphur dioxide gas, reaction may take place on the alkaline concrete surface. Dissolution and re-crystallization of the chemical products in the surface pores leads to disruption or 'fretting' of the surface. The underlying body of the concrete is normally unaffected.

- Loss of bond or cracks in ceramic tile coverings/brick slips — Although tiled façades and brick slip faced lintels are rarely the responsibility of an engineer during design, it is often to the engineer that others look for guidance when they fail. Their detachment may indicate failure of the adhesive or more deep-seated problems. Often the cause is failure to realize the different characteristics of concrete and facing material, for whereas concrete shrinks as it ages, ceramic materials (brick in particular) expands, and if joints are not provided at adequate intervals to accommodate the resulting movement, failure is inevitable.

3.9. Deposits on concrete surfaces

- Rust stains — Rust staining, together with fine cracks, is almost always an early sign of reinforcement corrosion. On the other hand, rust-like stains may also be the result of contamination of aggregates with pyrites (iron sulphide) which oxidizes on contact with air. Stain patches may also result from the rusting of nails and tie wires which have been left in the formwork during casting and rust deposits on the formwork before casting may also show as rust stains on the concrete surface. At construction joints, staining may result from unprotected reinforcement being left exposed for long periods of time between the casting of adjacent bays.
- Exudation — In alkali–silica reaction, highly alkaline solution in the cement can react with the silica content of some aggregates to produce a gel which will expand in the presence of water and cause hexagonal map-like cracking which breaks up the concrete mass. Viscous gel-like material may be observed to exude through pores or cracks.
- Efflorescence — Water moving through very permeable concrete can cause leaching of calcium hydroxide onto the surface which, in extreme cases, may form stalactites. Reaction with carbon dioxide in the air may cause the formation of a calcium carbonate deposit which is usually of whitish appearance (although sometimes stained by ground

water or rust) on the concrete surface. In general, efflorescence is unlikely to be harmful. On the other hand, excessive leaching of calcium hydroxide will increase the porosity of the concrete, making it more prone to chemical attack. Crystallization of other salts may also cause efflorescence.

- Dampness and leakage — Although it is common for water retaining structures to be constructed in reinforced concrete they are never completely watertight because concrete naturally absorbs and then, through evaporation, loses water. This is recognized by BS 8007 (3 in bibliography), which states that a water retaining structure can be considered watertight if the total drop in surface level of the contents does not exceed a specified figure over a given period of time. The extent of such absorption and leakage will depend on the mix proportions used and the quality of construction. Where there are imperfections in the concrete, such as cracks or voids, water leakage can be substantial.

There are many possible reasons for the occurrence of dampness and leakage in concrete building structures. The more common causes are

- cracks caused by shrinkage and restraint
- voids below the reinforcement due to plastic settlement
- voids below water bars at movement joints
- unsatisfactory construction joints
- failure to adequately grout up holes provided for formwork ties
- honeycombing or voids due to poor compaction or grout loss.

Identification of the causes of dampness and leakage is never quite straightforward as the place where liquid enters the structure is seldom where the dampness or leakage are apparent on the surface. Leakage could also occur as a result of many contributing causes. Ceramic tiles and other finishes on external walls often mask the true entry point for water.

3.10. Structural deformation

Failure to meet the specified limit states of ultimate strength or serviceability may be indicated by the following signs of distress

- excessive deflection
- excessive vibration
- buckling or distortion.

Well constructed reinforced concrete structures of conventional design should normally have a reasonable degree of redundancy to tolerate some deficiencies. It is however important to realize that a structure showing any of the signs listed above warrants immediate attention as they could represent early signs of pending failure. On the other hand, deformation is sometimes built in at the construction stage, for example, when insecure or inadequate formwork moves.

3.11. Construction defects

Fig. 3.2 is included to illustrate a number of construction defects — albeit in an unusually severe form. Due to lack of vibration concrete has become caught up on the reinforcement at the change of section. There has then been a significant time interval preceding placing of the next delivery of concrete, sufficient for the trapped concrete to gain enough strength to prevent the filling of the large void. At the base of the newer concrete the arching patterns of plastic settlement can be seen.

Although a number of defects already mentioned may originate from lack of care during construction those listed below are unlikely to result from any other cause.

- Honeycombing — occurs when concrete is not properly consolidated due either to inadequate vibration or poor design of formwork, allowing pockets of concrete to lie out of the reach of vibrators.
- Voids — usually the result of poor formwork design allowing pockets of trapped air to occur.
- Tearing — caused by a shock to the concrete some time after

Fig 3.2. An extreme example of construction defects

placing but before it has achieved adequate strength. Often the result of careless removal of formwork.

- Scouring — excess water rising in the forms can produce stream-like patterns in the surface of the concrete.
- Blowholes — small bubble-like holes in the surface, particularly where top formwork or sloping side forms have been used, caused by small pockets of air or water trapped by the forms.
- Powdered surfaces — caused by the use of new timber formwork made of wood containing retardant materials such as natural sugars.
- Incorrect cover — reinforcement may become displaced during construction if not adequately tied and supported. Although less cover is detrimental with regard to durability, excess cover, as can occur when top reinforcement in a slab is subjected to excessive trafficking during the placing of concrete, can reduce load carrying capacity.

3.12. *Construction features*

Although construction features such as construction and panel joints are not in themselves defects they can form locations where other defects are likely to become apparent. Shrinkage of a panel, for example, may cause a crack along the panel joint. The area around abrupt changes of section should always be critically examined.

3.13. *Common signs of distress in non-concrete components*

Signs of distress in non-concrete components may be manifestations of deflections, distortion, displacements or other movements in supporting concrete members. Such signs of distress would include the following

- cracks in in-fill walls
- cracks at corners of openings
- cracks between different materials, e.g. in-fill walls and concrete members
- distortion of door or window frames causing jamming
- distortion of members in partition walls.

Select bibliography

1. Allen R. T. L. *et al. The repair of concrete structures* Blackie, London, 1993, 2nd edn.
2. American Concrete Institute. *Guide for making a condition survey of concrete in service.* The Institution, 1993.
3. British Standards Institution. *Code of Practice for the design of reinforced concrete structures for retaining aqueous liquids*, BS 8007. BSI, London, 1987.
4. British Standards Institution. *Structural use of concrete*, BS 8110, Part 1 *Code of Practice for the design and construction*. BSI, London, 1985.
5. Building Research Establishment. *Floor screeds*, BRE, Garston, 1972, BRE Digest 104.
6. Building Research Establishment. *External rendered finishes.* BRE, Garston, 1976, BRE Digest 196.
7. Building Research Establishment. *Estimation of thermal and moisture movements of structures*, BRE, Garston, 1979, BRE Digests 227–229.

8. Building Research Establishment. *Fixings for non load-bearing cladding panels*, BRE, Garston, 1980, BRE Digest 235.
9. Building Research Establishment. *Damage to structures from ground-borne vibration*, BRE, Garston, 1990, BRE Digest 353.
10. Building Research Establishment. *Why do buildings crack?* BRE, Garston, 1991, BRE Digest 361.
11. Building Research Establishment. *Sulphate resistance of buried concrete*, BRE, Garston, 1992, BRE Report 164.
12. Building Research Establishment. *Sulphate and acid resistance of concrete in the ground*, BRE Garston, 1992, BRE Digest 363.
13. Building Research Establishment. *Effects of ASR on concrete foundations*, BRE, Garston, 1993, BRE IP 16/93.
14. Building Research Establishment. *Concrete: cracking and corrosion of reinforcement*, BRE, Garston, 1994, BRE Digest 389.
15. Building Research Establishment. *Assessment of existing high alumina cement concrete construction in the UK*, BRE, Garston, 1994, BRE Digest 392.
16. Bungay J. H. *Testing of concrete in structures*. Blackie, London, 1989, 2nd edn.
17. Campbell-Allen D. and Roper H. *Concrete structures: materials, maintenance and repair*. Longman, London, 1991.
18. Concrete Society. *Non-structural cracks in concrete*. The Society, London, 1982, Technical Report 22.
19. Gilbert R.I. *Time effects in concrete structures*. Elsevier, Oxford, 1988.
20. Holland R. *et al*. *Appraisal and repair of building structures — Introductory Guide*. Thomas Telford, London, 1992.
21. Institution of Structural Engineers. *Appraisal of existing structures*. The Institution, London, 1996.
22. Loss Prevention Council. *Property subject to structural movement: guidelines on the assessment of cracks*. The Council, London, 1995.
23. Mays G. (ed.). *Durability of concrete structures — investigation, repair, protection*. Spon, London, 1992.
24. Perkins P. H. *Repair, protection and waterproofing of concrete structures*. Elsevier, London, 1986.
25. Puller-Strecker P. *Corrosion damaged concrete — assessment and repair*. Butterworth, Oxford, 1987.
26. Rainger P. *Mitchell's movement control in the fabric of buildings*. Batsford, London, 1983.

4

Principal investigation

4.1. Introduction

Basic detective work was outlined in Chapter 2 — the initial appraisal. This chapter will deal with management of the principal investigation and the means by which more extensive data can most effectively be gathered. The assessment of the strength of the concrete structure, however, will be left to be dealt with in Chapter 6.

4.2. Preliminary work

During the initial appraisal hypotheses regarding the possible causes of defects then observed are likely to have been postulated, each of which requires examination in detail during the principal investigation, perhaps to confirm the diagnosis but certainly to identify the scale of the problem in terms of both present extent and future threat.

It may well be worth carrying out a literature search or seeking the views of specialist firms or laboratories at this stage in order to locate instances where similar defects have occurred. A few hours spent browsing in a good technical library, such as those of the Institution of Civil Engineers or the Institution of Structural Engineers, may well provide the lead which the investigator is seeking. Conferences covering the general subject of appraisal and repair are held from time to time and their proceedings contain many case studies. It is foolish to think that the problem under consideration is unique; conversely it is dangerous to accept uncritically another investigator's conclusions without thoroughly checking both the credentials and motivation of the investigator and the significance of the apparent similarities with one's own problem.

4.3. The nature of the evidence

The use to which data is to be put — and the previous experience of the investigator — should determine, to some extent, the means by which it ought to be acquired. For example, it may be prudent to employ recognized specialist firms to carry out some parts of an investigation if the results are required to stand up to rigorous legal scrutiny.

4.4. The extent of investigations

Investigation is expensive — particularly where access is difficult. The aim should therefore be to collect only that data which is necessary; the amount required will depend upon why the data is needed.

The principal reasons for requiring data are

- to dispel uncertainty or ignorance regarding the physical condition of hidden constructions
- to determine the physical extent of defective areas
- to predict, where defects or deformations are extensive, the residual strength of a structure
- to test hypotheses, eliminate the invalid and confirm the valid
- to provide a basis for establishing appropriate remedial action
- to provide a basis for estimating the cost of remedial work.

One critical method of deciding the extent of investigation required — based upon the Pareto principle of diminishing returns — is to establish what the cost of not carrying out any part of the investigation might be. For example if, in a 5% random sample of the beams in a large office block, one in eight were found to be of unacceptable strength it would, in order to provide more accurate cost predictions, be worth testing a larger sample before drawing up a contract for strengthening works. Conversely, if six out of eight beams had been found to be unacceptable there would be a strong argument for any further testing, if judged to be necessary at all, to be left to form part of the remedial contract.

4.5. Organizing the principal investigation

As with the initial appraisal it is important to draw up a field specification (see Section 2.2 above) and to use proformas and location drawings to describe and locate findings. The danger of repeating what has already been carried out can be obviated if the investigators are given copies of previously completed proformas and sketches — the latter especially, as their preparation can be very time consuming. There will also be instances where signs of distress noted during the initial appraisal are sufficient to determine the extent of deterioration and the cause of the defects, making a more detailed survey unnecessary.

The following notes give guidance on methods of data collection for each of the most common concrete problems. Although the tools provided for the initial appraisal will again prove useful there are several data collection techniques which can speed up the operation and these are described below. To assist comparison with Chapter 3 the subject matter is treated in a similar sequence.

4.6. Measurement and description of defects in general

Essential data relating to all defects is as follows.

- Location, type (e.g. crack, spalling, dampness, etc.) and dimensions of the structural member in which each defect or group of defects occur, noting any change in geometry of the member near to the defects.
- Dimensional relationship of defects to reinforcement, movement or construction joints, repairs, modifications or any other features which could help to establish causes. A cover meter should be used to locate the position of reinforcement.

- Whether, and to what extent and in what form (and colour if relevant), any other defects are associated with those being described.
- The environment in which the defects have occurred (e.g. external and open to rain, buried in waterlogged soil, in centrally heated building, above chlorinated water in swimming bath, etc.). Always bear in mind that environmental conditions may have changed since the damage was done. Where temperature and humidity may be critical, data recording devices are available.

A number of techniques have been developed for testing concrete in-situ and these are summarized in Table 4.1.

The reference work for most tests on concrete is British Standard 1881: *Testing concrete*. The bibliography to this chapter contains references to those Parts of the Standard likely to be of most use to the investigating engineer.

A critical appraisal of all the tests enumerated in Table 4.1, and many other tests, is given in J. H. Bungay's *Testing of concrete structures* (21 in bibliography).

4.7. Cracking

Additional data required regarding cracking is as follows.

- Shape of each crack or group of cracks (best in the form of a sketch or photograph of each face on which they are seen) and whether any cracks reach the edge of the structural member in which they occur.
- Depth and profile (e.g. tapered, straight, irregular, etc.) of each crack and, in particular, whether it passes completely through the structure. In some cases, however, it will be extremely difficult, and probably unnecessary, to establish the depth of a crack. Only if other data is insufficient to support an adequate diagnosis for appropriate remedial works to be decided upon should this be pursued further. For larger cracks an endoscope may be a useful probing device and for smaller cracks a hole could be drilled, approximately in line with the crack, to accept the probe of an endoscope. Ultrasonic pulse-velocity testing may also be useful — see Section 4.17 below.

Table 4.1 Principal in-situ testing techniques

Use	Technique	Observations
Locating voids and delaminations	Hammer testing	Effective to about 100 mm
Concrete quality, locating voids and cracks	Ultrasonic pulse Velocity testing	Large areas covered rapidly
Examination of voids and cracks	Endoscope survey	
Concrete strength	Schmidt hammer Windsor probe	Indicative only of strength in cover concrete
Concrete strength	Laboratory testing of cores	
Likelihood of corrosion of reinforcement	Half-cell potential mapping	Large areas covered rapidly
	Resistivity measurement	Measures electrical resistance of concrete cover
Depth of carbonation	Site phenolphthalein test	Requires clean broken surface
Chloride contamination	Site tests on concrete powder using Quantab strips	
Sulphates contamination	Laboratory tests on drilled samples	
Depth of cover Size of reinforcement	Cover meter survey	
Strength of element or structure	Load testing	Expensive
Permeability/water absorption	ISAT (see Section 4.18)	In-situ test
Density, permeability Water absorption Chemical analysis Type of cement/ replacement materials Density	Laboratory tests on broken out samples	
Aggregate type	Stereoscopic examination of broken out samples	
Aggregate, cement type	Petrographic analysis of broken out samples	

Removal of plasterboard lining disclosed this advanced deterioration

- Width. A measuring device consisting of plain black lines of different thicknesses on a light or transparent background which can be laid over cracks will speed up assessment of widths.
- Alignment of surfaces either side of each crack to establish whether, by how much and in which directions, one side of the crack has moved or is moving in relation to the other. This information is required in three dimensions and can best be obtained using a steel square, a steel straight edge and a soft pencil as illustrated in Fig. 4.1.

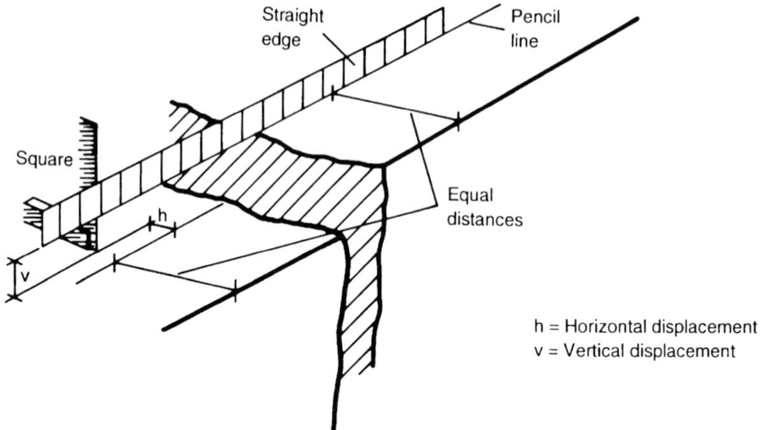

Fig. 4.1. How displacements across cracks may be measured

- Movement across cracks with time may be important, particularly in predicting future possible degradation. Guidance on monitoring is discussed in the Introductory Guide to this series (27 in bibliography).

4.8. *Surface deterioration*

Additional data required regarding surface deterioration is as follows.

- Shape of each item of spalling, delamination or weathering (best in the form of a sketch or photograph of each face on which the defect is seen). Tapping with a hammer should identify the boundaries of delamination but, on the other hand, could dislodge it. However if this happens it would have been in a dangerous condition anyway — the important thing is to be aware and to take precautions!
- Depth and profile of defect in general terms.
- Samples of any exuded gel should certainly be taken for laboratory confirmation of suspected alkali–silica reaction. It may be wise also to take samples of efflorescence for laboratory examination.

4.9. *Surface deposits*

Additional data required regarding surface deposits is as follows.

- Shape of each item of efflorescence, rust staining or dampness (in the form of a sketch or photograph) and a description of the colour of rust staining (dark brown could indicate a more serious situation than light brown) or efflorescence; the colour and smell of leakage or patches of dampness (mould growing on damp areas could be a threat to timber or public health and colour and smell might help to indicate the source or route of the liquid).
- Dimensional relationship of the defect to water pipes, sewers or drains, or any other features which could help to establish cause.

4.10. *Deformations*

Additional data required regarding deformations is as follows.

- The dimensions of deflections, buckling or distortions — if a suitable plane of reference can be provided it may be possible to use photography to record dimensions.
- Evidence to establish whether deformations are the result of events which took place during construction or are due to subsequent events.
- A description of vibrations by reference to their frequency, magnitude and trigger mechanisms.

4.11. *Construction defects*

Additional data required regarding construction defects is as follows.

- For the essential data requirements relating to some commonly occurring construction defects (cracks which originated before the concrete hardened and deformation due to construction errors and mishaps) reference should be made to the foregoing paragraphs.
- Size of each item of honeycombing, voids, or scoring if they are

to be repaired either for aesthetic reasons or to improve durability.

4.12. Reinforcement corrosion

Severely corroded bars should be exposed and the proportion of metal which has been removed by corrosive action assessed directly. Consideration must then be given to the effect the loss of section is likely to have on the strength of the affected member and its capacity to perform its structural function satisfactorily.

The extent of corrosion of reinforcement hidden within apparently sound, or almost sound concrete can be indicated by a combination of two methods — half cell potential mapping and resistivity testing of the concrete. Both of these methods are described in the CIRIA report prepared by Peter Puller Strecker (31 in bibliography) and further details, including their limitations, are given in *The Durability of Concrete Structures* edited by Geoff Mays (29 in bibliography). A number of commercial investigation companies have developed devices for carrying out mapping of large areas using the half-cell potential method.

These techniques provide the investigator with maps of the concrete surface investigated on which contours represent the degree of reinforcement corrosion as indicated by the instrumentation used. The maps can be used to differentiate between areas where repair is required and those where protection would be a cost-effective alternative. It is important however to realise that such measures can only give an indication of the likely extent of corrosion, and confirmation of the actual extent during remedial work, by selected breaking out of the concrete cover, would be advisable.

4.13. Depth of carbonation

Only fresh fracture surfaces should be used for carbonation testing and drilled surfaces avoided. Such a surface can be obtained by drilling a number of holes in the concrete and breaking out the portion of concrete between them. The fracture surface is then sprayed with a phenolphthalein indicator which develops a red colour at pH values indicative of carbonation having taken place.

The technique is described in detail in the BRE Information Paper 6/81 (20 in bibliography).

4.14. *Chlorides*

Samples of concrete obtained by drilling are sufficient for a laboratory test to determine chloride content. To establish whether chlorides are widespread or restricted only to the surface layers of the concrete, samples from different locations and depths should be taken. Sampling by drilling is described in detail in BRE Information Paper IP 13/77 (19 in bibliography). An alternative method which can be used on site is the commercially available 'Quantab' test strip. This is a small plastic strip which, when inserted in a solution of concrete powder, changes colour to indicate the proportion of chlorides present.

4.15. *Depth of cover to reinforcement*

Modern electronic covermeters can indicate the depth of cover to an accuracy of about 5 mm after site calibration but small bars near the surface may be confused with large bars with greater cover. British Standard 1881 part 204 (13 in bibliography) gives detailed guidance on the use of covermeters. It should be remembered that some stainless steels are not magnetic and may not therefore be located by cover meters using magnetic detection techniques. Meters using other techniques are however available commercially.

4.16. *In-situ strength of concrete*

Although the testing of core samples is the most accurate method of determining the in-situ compressive strength of concrete it is both expensive and disruptive. A spring loaded surface rebound device known as a Schmidt hammer can be placed against the concrete and, when the energy is released, the rebound is measured on a scale incorporated with the device. This energy is proportional to the compressive strength of the concrete tested and the device is so simple to use that several tests can be carried out in less than a minute. The use of the Schmidt hammer will allow areas of weakness to be determined but is not sufficiently reliable to predict available concrete strength for use in design calculations, particularly of concrete well below the surface.

Several alternative site tests to determine the strength of concrete are available and the merits of these are discussed in J. H. Bungay's *Testing of concrete structures* (21 in bibliography).

4.17. Hidden defects

Although voids, hidden cracks and incipient delamination may be shown up by the Schmidt hammer test it is by no means easy to differentiate them from each other or from weak areas of concrete.

A more satisfactory method is to use ultrasonic pulse velocity (UPV) testing which is described in some detail in *The Durability of Concrete Structures, Investigation, Repair and Protection*, edited by Geoff Mays (29 in bibliography). This technique is however only effective where both sides of a member are accessible and would therefore be of limited use for a ground floor slab or retaining wall. Care has to be taken to avoid confusing the evidence for voids with that for reinforcement.

4.18. Other properties of concrete

Water absorption and permeability tests can be useful indicators of how readily the agents which cause defects can progress through the concrete. Such tests can also be used to provide evidence which might help to explain why damp patches occur in some locations and not others. British Standard BS 1881: Part 208 (16 in bibliography) specifies an in-situ test for surface absorption (ISAT) and this is discussed in *Concrete Structures, Investigation, Repair and Protection* edited by Geoff Mays (29 in bibliography).

A number of useful laboratory tests are also described in the latter publication including

- stereoscopic examination of concrete
- density and excess voidage
- chemical analysis
- petrographic analysis of concrete and aggregates
- cement type and replacement materials.

4.19. Size of reinforcing bars

Some designs of covermeters can be used to establish the size of reinforcing bars in a concrete element. However, because of the need to distinguish between single large bars and bundles of smaller bars and the possibility of laps occurring where readings are taken, some experience in their use is required before a survey is undertaken. It is recommended that callibration is undertaken by breaking out the reinforcement in selected areas after the covermeter survey. British Standard 1881 Part 204 (13 in bibliography) gives detailed guidance on the use of covermeters and the subject is discussed in detail in J. H. Bungay's *Testing of concrete in structures* (21 in bibliography).

Select bibliography

1. Addleson L. *Building failure: a guide to diagnosis, remedy and prevention*. Butterworths, Oxford, 1992.
2. Allen R. T. L. *et al*. *The repair of concrete structures*. Blackie, London, 1993, 2nd edn.
3. American Concrete Institute. *Guide for making a condition survey of concrete in service*. The Institute, 1993.
4. American Concrete Institute. *Manual of concrete inspection*, ACI Special Publication 2, The Institute, 1992, 8th edn.
5. British Standards Institution. *Testing concrete*, BS 1881, Part 5: *Methods for testing hardened concrete for other than strength*. BSI, London, 1970.
6. British Standards Institution. *Testing concrete*, BS 1881, Part 114: *Methods for determination of density of hardened concrete*. BSI, London, 1983.
7. British Standards Institution. *Testing concrete*, BS 1881, Part 120: *Method for determination of the compressive strength of concrete cores*. BSI, London, 1983.
8. British Standards Institution. *Testing concrete*, BS 1881, Part 122: *Method for determination of water absorption*. BSI, London, 1983.
9. British Standards Institution. *Testing concrete*, BS 1881, Part 124: *Analysis of hardened concrete*. BSI, London, 1988.
10. British Standards Institution. *Testing concrete*, BS 1881, Part 201: *Guide to the use of non-destructive tests for hardened concrete*. BSI, London, 1986.

11. British Standards Institution. *Testing concrete*, BS 1881, Part 202: *Recommendations for surface hardness testing by rebound hammer*. BSI, London, 1986.
12. British Standards Institution. *Testing concrete*, BS 1881, Part 203: *Recommendations for measurement of velocity of ultrasonic pulses in concrete*. BSI, London, 1974.
13. British Standards Institution. *Testing concrete*, BS 1881, Part 204: *Recommendations for the use of electromagnetic covermeters*. BSI, London, 1988.
14. British Standards Institution. *Testing concrete*, BS 1881, Part 205: *Radiographic tests*. BSI, London, 1986.
15. British Standards Institution. *Testing concrete*, BS 1881, Part 208: *Near to surface tests*. BSI, London, 1992.
16. British Standards Institution. *Guide to the assessment of concrete strength in existing structures*, BS 6089. BSI, London, 1981.
17. Building Research Establishment. *Assessment of existing high alumina cement concrete construction in the UK*. BRE, Garston, 1994, BRE Digest 392.
18. Building Research Establishment. *Simplified method for the detection and determination of chloride content in hardened Portland Cement concrete*. BRE, London, 1977, BRE Information Paper IP 12/77.
19. Building Research Establishment. *Determination of chloride and cement content in hardened Portland Cement concrete*. BRE, London, 1977, BRE Information Paper IP 13/77.
20. Building Research Establishment. *Carbonation of concrete made with dense natural aggregates*. BRE, London, 1981, BRE Information Paper IP 6/81.
21. Bungay J. H. *Testing of concrete in structures*. Blackie, London, 1989, 2nd edn.
22. Chess P. and Grønvold F. *Corrosion investigation: a guide to half cell mapping*. Thomas Telford, London, 1996.
23. Concrete Society. *Non-structural cracks in concrete*. The Society, London, 1982, Technical Report 22.
24. Concrete Society. *Permeability testing of concrete: a review of methods and experience*. The Society, London, 1988, Technical Report 31.
25. Concrete Society. *Assessment and repair of fire-damaged concrete structures*. The Society, London, 1990, Technical Report 33.
26. Concrete Society. *Sub-surface radar surveying*. The Society, London, 1994, CS 109.
27. Holland R. *et al*. *Appraisal and repair of building structures — Introductory Guide*. Thomas Telford, London, 1992.

28. Institution of Structural Engineers. *Guide to surveys and inspections of buildings and similar structures*. The Institution, London, 1991.
29. Mays G. (ed.). *Durability of concrete structures — investigation, repair, protection*. Spon, London, 1992.
30. Perkins P. H. *Repair, protection and waterproofing of concrete structures*. Elsevier, London, 1986.
31. Puller-Strecker P. *Corrosion damaged concrete — assessment and repair*. Butterworth, London, 1987.
32. Rainger P. *Mitchell's movement control in the fabric of buildings*. Batsford, London, 1983.

5

Causes of defects

5.1. Identifying the causes of defects

Tempting though it is to arrive as quickly as possible at a solution to a deterioration problem it is more important to ensure that all the physical and test evidence supports the hypothesis which is eventually chosen. It is not uncommon for an engineer to be asked to advise on a structure which has already been 'repaired'. The identification of the cause or causes is a critical process requiring an open but questioning mind. The process concludes only when the evidence from site investigation and results from site and laboratory tests explain and reinforce one another.

The engineer carrying out the appraisal should seek to develop a sound understanding of the various mechanisms that can result in defects in concrete structures if causes are to be correctly diagnosed, evaluation of the structural adequacy and of the residual service life of the structure under investigation properly assessed and appropriate repair strategies selected.

The mechanisms of decay associated with the various defects which afflict reinforced concrete structures have been fairly well researched and documented. The purpose of this chapter is to help the appraising engineer by categorizing the more common causes of defects and by providing concise explanations of the mechanisms involved. References are given to other literature where more detailed information may be found. It should be borne in mind that defects may appear as a result of a combination of causes.

5.2. *Quality of concrete*

A factor which will affect, in different ways, almost all the defects likely to be encountered in reinforced concrete is the quality of the concrete itself. For example, minimum cement content was not a requirement of Codes of Practice until the 1970s as adequate strength was, before that date, considered to equate with adequate durability. The permeability of concrete is of considerable importance in determining how quickly harmful gasses, such as carbon dioxide, will permeate the cover to reinforcement or if it will permeate at all. It follows that the investigating engineer will wish to know as much as possible about the quality of the concrete he is dealing with before drawing conclusions from other evidence and certainly before making any recommendations for repair.

5.3. *Categorization of common causes of defects*

This chapter will, as far as possible, follow a sequence similar to that used in the two preceding chapters. To further assist the reader, the causes are first listed and then dealt with in the order in which they appear in the list.

Causes of defects in reinforced concrete building structures may then be categorized as follows.

- Effects due to restraint to volume changes in concrete
    - plastic settlement
    - plastic shrinkage
    - early thermal contraction
    - long term drying shrinkage
    - shrinkable aggregates
    - surface crazing
    - thermal movement.
- Effects related to corrosion
    - corrosion of reinforcement
    - carbonation of concrete cover

Prising open a crack in this column revealed significant corrosion

- chloride contamination of concrete.

- Effects due to other chemical changes within the concrete
  - alkali–silica reaction
  - sulphates attack
  - acid attack
  - efflorescence
  - lime bloom
  - atmospheric pollution
  - deterioration of high alumina cement (HAC) concrete.

- Effects related to inadequate strength or robustness
  - design deficiencies
  - construction deficiencies
  - foundation deficiencies
  - maintenance deficiencies
  - misuse.
- Effects related to environmental causes
  - climatic effects
  - ground movement
  - water and dampness
  - fire.

5.4. Plastic settlement

Within setting concrete not only are chemical reactions taking place but the mass of concrete starts to behave as an increasingly stiff plastic material having, in the first few hours, little strength. Under the force of gravity the larger stones in a mix of high water content will flow downwards, displacing cement paste and sand. At the same time less dense pockets of concrete may continue to consolidate, particularly in those parts of a structural member which have significantly greater depth than surrounding parts. Furthermore voids may form due to leakage of cement paste through badly constructed formwork and construction joints. In attempting to fill the voids thus created the concrete, which has by this time become too stiff to flow as a liquid, tears itself upon any obstruction, such as a reinforcing bar or a projection or corner in the formwork, which impedes its plastic flow.

Although essentially due to construction malpractice, there are situations where designers' details (e.g. excessive quantities of reinforcement in narrow walls or columns, abrupt changes of section, insufficient thought given to the design of prestressing ducts and water bars) make it difficult for the problem to be entirely eliminated by the construction team alone.

5.5. *Plastic shrinkage*

If one considers any concrete section it should be obvious to an engineer that strains will be set up if the outer layers are allowed to dry out at a rate vastly different to that at the heart of the section. Designers usually specify curing regimes in an attempt to prevent surfaces drying out too quickly but, because the weather in the United Kingdom has many (humid, overcast, windless) days when the specified precautions may be unnecessary, construction teams can become forgetful of their duties with regard to curing. Plastic shrinkage cracking (or surface crazing — see below) is the inevitable result if a curing regime appropriate to the wind on the day of placement is not used. Obviously temperature, exposure to the direct rays of the sun and lack of humidity are the more obvious causes but most construction teams seem to be aware of the dangers of these phenomena.

5.6. *Early thermal contraction*

Thermal cracking in thick concrete members within the first few weeks following placement is caused by steep temperature gradients across the section which prevent the concrete from setting at a constant rate throughout. These gradients are caused by the heat generated during the setting process (the heat of hydration) being unable to escape from the centre at the same rate as it does from the surrounding concrete. The phenomenon can also occur when a thin wall is cast against a larger concrete base. Much advice has been given to designers and construction teams over recent years to help them avoid this situation occurring.

5.7. *Long term drying shrinkage*

As concrete ages excess water will evaporate from its surfaces over a period of time, the length of which will depend upon the thickness of the section and environmental conditions. Loss of this water will lead to shrinkage of the concrete. In thin concrete sections the internal strains caused if the shrinkage cannot be accommodated by movement may be critical. This is a perfectly natural process and, if the recommendations of codes of practice are followed, should not cause problems. High water–cement ratios, excessive distances between construction and movement joints, inappropriate choice of the sequence of casting adjacent bays and anchorages which prevent movement taking place as planned are all possible causes of the defects commonly associated with long term drying shrinkage.

5.8. *Shrinkable aggregates*

The shrinkage of concrete containing certain aggregates (primarily certain forms of dolerite) may be three times greater, over a period of up to twelve months, than that made with other aggregates. BRE Digest 35 (3 in bibliography) gives some guidance on the identification of such aggregates.

5.9. *Surface crazing*

Surface crazing, like plastic shrinkage cracks (see Section 5.5 above), is caused by inadequate curing regimes during construction.

5.10. *Thermal movement*

It is said that a contributory cause for the inclination of the Campanile at Pisa is the twisting motion induced in it by the difference in temperature between the shaded and un-shaded surfaces as the day progresses. The same phenomena has been observed on tall reinforced-concrete structures. Differences in temperature between the seasons of the year also induce large slabs of concrete to grow and shrink significantly and many locations near industrial processes involve significant temperatures.

5.11. *Corrosionmechanisms*

By far the greatest number of the defects which afflict reinforced concrete are associated with the corrosion of reinforcement. In a book entitled *Cathodic Protection of Reinforcement Steel in Concrete* by K. C. G. Berkeley and S. Pathmanaban (1 in bibliography), the authors describe five different forms of corrosion from which metals may suffer. Four of the five mechanisms rarely occur in reinforced concrete structures and, in the following notes, are only briefly described. Unless the reader bears in mind that corrosion cannot take place unless moisture is present the following may make disquieting reading!

- Concentration cell corrosion — This occurs when the same piece of metal is exposed to differing electrical potentials along its length, as is often the case with reinforcement in concrete where significant differences in chemical balance can occur along the length of a bar. Discussion of this topic continues at Section 5.12 below.

- Differential-aeration cell corrosion — This is a specific case of concentration cell corrosion and occurs when a piece of metal is exposed to differing concentrations of dissolved oxygen along its length such as might occur when part of a structure is totally immersed — (the anode) — and part — (the cathode) — is in a more active environment such as the splash zone or close to the turbulent conditions often associated with an inlet or outlet for water.

- Galvanic cell corrosion — This requires two dissimilar metals to be in electrical contact which could occur through the medium of damp or wet concrete. All metals have a galvanic relationship to one other so that in any pair one will always form the cathode and other the anode. Berkeley and Pathmanaban (1 in bibliography) list the most common metals showing which would be anodic and which cathodic in relationship to carbon steel in sea water at 25 °C. From this it can be seen that aluminium and zinc would be anodic (i.e. protective) in relation to carbon steel but most other metals (see below) would be cathodic and therefore likely to cause the steel to corrode. The type of situations where galvanic cell corrosion of reinforcement in concrete could occur is where fittings holding copper or brass pipework onto the face of a reinforced concrete structure come into contact with the steel reinforcement, perhaps, through their fixings.

  The following list gives the order of nobility of the metals which are detrimental to mild steel — those placed highest on the list being the most detrimental

- gold
- silver
- copper
- brass
- bronze (including phosphor-bronze and gunmetal)
- lead
- stainless steel
- cast iron.

- Stray current corrosion (also known as induced cell corrosion) — as the name suggests, strong electrical current from a variety of sources can cause steel in reinforced concrete to corrode. Berkeley and Pathmanaban (1 in bibliography) list variations in the magnetic field of the earth, extra-high-tension power lines, DC tram or railway systems or adjoining but unrelated cathodic protection systems. Bases of electrical transformers might provide another site for this form of corrosion.
- Bacterial corrosion — Unlike other corrosion mechanisms, this does not require the presence of oxygen — indeed the reaction will not take place in oxygen-rich situations. The corroding bacteria are found primarily in clay and silty soils but could occur in farm buildings if organic matter is allowed to decay. As the bacteria convert steel into its sulphide form it is relatively easy, from laboratory samples of the rust, to differentiate from other forms of corrosion.

5.12. Corrosion of reinforcement

Except for bacterial corrosion (see above) certain factors are necessary before the steel reinforcement in concrete will corrode and these are

- water — or, at least, dampness
- oxygen
- destruction of the passivation layer that protects the steel from

The disruptive power of reinforcement corrosion

corrosion, which may be caused by carbonation of the concrete or by chemical imbalance due to the presence of chlorides.

The passivation layer

The principal protection of steel in concrete is a thin layer of normally insoluble material which forms as a result of reaction between the steel and the surrounding concrete and which is known as the passivation layer. The steel, so protected, is said to be 'passivated'. The conditions which permit this layer to remain stable, and therefore protective, depend upon the electrical

Table 5.1 Minimum pH values required to maintain passivity — concrete without significant chloride level

Electrical potential	+800 mV	+400 mV	0	−400 mV	−800 mV	−1200 mV
pH	3.5	5	6.5	8	9.5	immune

potential and alkalinity of the concrete. Table 5.1 shows the generally accepted pH values below which breakdown is likely to occur.

It should be noted however that in highly alkaline conditions (i.e. above a pH of 12) the passivation layer may also be vulnerable in conditions of low electrical potential.

Cover

The importance of adequate cover in protecting the passivation layer from disruption cannot be over-stated. The requirements of modern codes of practice in respect of thickness of cover are adequate only if the cover is of dense well compacted concrete having the cement content specified, otherwise the concrete will carbonate (see Section 5.14 below) and so lose its protective capacity. The same rules apply to repairs except where epoxy resin mortars are used. Epoxy resins work on an entirely different principle in providing protection to the steel (see Chapter 8).

5.13. Corrosion damage to concrete

Corrosion of steel is an expansive process capable of generating forces strong enough, in time, to break open the surrounding concrete. This not only removes the protection provided by the concrete but can seriously reduce the load carrying capacity of a structural member. Even before large scale surface cracking is apparent at the surface there can be some reduction in the bond between steel and concrete.

5.14. Carbonation

Carbonation is a natural chemical process during which the calcium hydroxide content of the concrete is converted to calcium carbonate by the action of dissolved carbon dioxide. Apart from the outer few millimetres, dense concrete is normally very resistant to carbonation but the less well-compacted locations (e.g. corners of columns) are likely to be carbonated to a greater depth than is the case elsewhere. The effect of carbonation which concerns us is the reduction in alkalinity which accompanies the process, as it is enough to destroy the passivation layer (see Section 5.12 above) which in turn increases the vulnerability of the reinforcement to corrosion. On the other hand, fully carbonated concrete is both stronger than non-carbonated concrete and more resistant to the passage of a number of harmful substances. In a relatively dry situation (e.g. inside a building where condensation on the concrete surface is minimal) carbonation may never lead to corrosion of the reinforcement — even though, in such conditions, the carbonation front itself may proceed very rapidly.

5.15. *Chloride contamination*

Calcium chloride was once used to speed up the setting of concrete. Other chlorides (e.g. road salt — sodium chloride) may find their way into reinforced concrete. While they have little effect on the concrete itself they can cause rapid corrosion of reinforcement where there is moisture and oxygen available. Even condensation (such as may be caused on the underside of an existing, poorly insulated, concrete roof slab by the addition of a false ceiling) may be sufficient to create the conditions conducive to corrosion, where chlorides contained in the concrete had not given trouble before. Some industrial processes may produce corroding chemicals and even unlikely materials such as sugar can be troublesome.

5.16. *Alkali–silica reaction*

Cements produced in some works contain more alkali than those from others. Some aggregates contain minerals (certain silicates) which are not tolerant of this higher alkalinity. Put the two together in a concrete and the non-tolerant minerals will produce an

Total breakdown of a reinforcing bar — aided by chlorides in the concrete

expanding gel which finally ruptures the concrete. Moisture is required for the reaction to take place. Most of the problems in the United Kingdom have occurred in the South West but the terrace gravels of the Trent and Thames valleys contain silicates which could prove troublesome in the future. Although the process is now understood, there is as yet no treatment which can be universally recommended. There is however a standing committee which keeps the matter under review. Further advice is contained in the Concrete Society's Technical Report 104 and the Institution of Structural Engineers' publication of 1992 (7 and 9 in bibliography).

5.17. *Sulphates attack*

Sulphates are present in cement to control the rate of setting and can also be found in brackish water or trade effluent. In rare cases unscrupulous operators have been known to add gypsum plaster to improve the setting time of concrete. Their principal source, however, is in the ground in varying amounts, and there are precautions (e.g. the use of sulphate-resistant cement) which specifiers can select which are appropriate to the local concentration. Where such precautions are not taken, the sulphates may react with the tricalcium aluminate in the concrete to form ettringite. The process is expansive and leads eventually to the complete disintegration of the concrete.

5.18. *Acid attack*

Being an alkaline material, concrete is particularly susceptible to acid attack which usually occurs in an industrial situation or due to spillage or acidic ground water but can also be caused by atmospheric pollution (see below) or by building on a contaminated site without taking adequate precautions. Other sources of acid could be milk or fruit juices.

5.19. *Attack by other aggressive materials*

In addition to the aggressive materials already referred to there are a number of chemical compounds which may be found in buildings and which can attack concrete. These include the following

- ammonium compounds
- caustic soda
- water if of low hardness or low alkalinity.

5.20. *Efflorescence*

Efflorescence is caused by chemical salts passing through the concrete and being left on the surface when the water which carried them has evaporated.

5.21. *Lime bloom*

If water flows through porous concrete or through joints in the concrete it often dissolves the calcium hydroxide in the concrete and deposits it as layers (sometimes also as stalactites and stalagmites) on the concrete surface. The unsightly and varied colour of lime bloom originates from the ground through which the water may have passed.

5.22. *Atmospheric pollution*

In addition to dirt, which can adhere to concrete surfaces and be deposited in unsightly concentrations by rainwater run-off, algae and lichen growths can establish themselves in damp and shaded situations, particularly where the concrete is so porous that it will hold water over the long periods necessary to sustain them. Over large parts of the country, but particularly in industrial areas and large conurbations, sulphur dioxide gas, which is acidic, will react with the alkaline surface of concrete causing disruption. This can leave the surface susceptible to frost damage and if the concrete contains porous aggregates, deeper seated problems may be caused.

5.23. *High alumina cement concrete*

High alumina cement (HAC) concrete is particularly useful at temperatures above 100 °C. It has however, in the past, been used for the manufacture of precast building components without any allowance being made for the loss of strength which can take place during its life due to a chemical change in the concrete. In the 1970s there was a sustained programme of searching for HAC concrete components but there may still be a few which have not yet been located.

The chemical change is very slow in cold conditions and does not occur at all in completely dry situations. It will however occur in the warm humid conditions within a normal building. HAC concrete made with a low water–cement ratio and used in small sections may not lose a great deal of strength as the heat generated during setting, and the free water then available, are critical.

5.24. Design deficiencies

Design errors

- Incomplete understanding of the possible modes of failure and structural behaviour resulting in inadequate provisions of restraints and ties for overall stability and robustness of the structure. This can also result from a tendency towards 'tunnel vision' by designers, who get over-concerned with design for strength alone and overlook the overall aspects of stability and robustness.
- Errors in design assumptions and concepts resulting in inappropriate modelling of the structure.
- Failure to recognize the limitations of analytical methods and computer software resulting in the use of inappropriate methods of analysis.
- Incorrect interpretation of code requirements or failure to recognize the conditions underlying code provision.
- Incorrect assessment of loadings, e.g. dead, imposed, wind and accidental loads such as earthquake and blast.
- Failure to recognize and cater for secondary effects.
- Incorrect assessment and inadequate provisions for environmental actions and micro-climates in design for durability.
- Incorrect assessment and inadequate provisions for movements, e.g. thermal movement, frost heave, shrinkable clay, changes in ground water level, differential settlement of foundations or creep.
- Failure to recognize the effects of restraints on movements.
- Failure to exercise critical checks on manual or computer design outputs. In particular, acceptance of computer output without any critical appraisal to see whether the results do actually make sense.
- Lack of coordination or of a critical overall check when designs of foundations, substructure, floor slabs, shear walls

or special structures are designed by specialists outside the main design team.

- Failure to give clear and specific instructions for the detailing of the design and the drawing up of contract specifications.
- Computational errors.
- Failure to exercise proper quality control procedures during design.

Detailing errors

- Errors and omissions in the transfer of design requirements and outputs to drawings.
- Inadequate understanding of design assumptions and requirements.
- Insufficient attention to practical aspects of construction in the provision of details (e.g. congestion of reinforcement, connection details).
- Inadequate provision of details, e.g. anchorages, laps and curtailment of reinforcement.
- Inadequate attention to points of stress concentration, e.g. sudden change of sections at corners of openings.
- Failure to exercise proper and critical checks on manual or computer drawing outputs.
- Computational errors.

5.25. *Construction deficiencies*

- Departures from design details.
- Departures from material specifications (e.g. inadequate cement content or excessive water content).
- Poor workmanship, resulting in
    - inaccurate quantities of ingredients (e.g. inadequate cement, excessive water, excessive additives)

  - incorrect sizes of sections, incorrect sizes and positions of reinforcement
  - insufficient or excessive concrete cover thickness
  - inadequate or excessive compaction of concrete
  - inadequate curing
  - inclusion of foreign matter in the concrete (e.g. boxing-out material).
- Inappropriate provision for and preparation of construction joints.
- Failure to conduct the specified tests.
- Failure to exercise proper quality control procedures during construction.

5.26. *Foundation deficiencies*

Inadequate foundations, although they may arise from any of the causes listed in Sections 5.24 and 5.25 above and from deterioration due to decay mechanisms such as sulphates attack (Section 5.17), can also arise because designers fail to understand the nature of the ground under the foundations.

5.27. *Maintenance deficiencies*

- Lack of periodic inspection due to the misconception that concrete structures will last forever and are maintenance free.
- Lack of a routine maintenance programme believing that minor defects such as corrosion of reinforcement or cracks are of no immediate consequence to the occupiers and therefore remain as a low priority in the owners' overall business plan. This is the result of a failure to understand that the remaining service life of a concrete structure depends largely on the recognition and rectification of early signs of distress. There is also a failure to appreciate the enormous repair cost that can result if such defects are allowed to deteriorate further.
- The use of inappropriate repair methods and materials without appropriate professional advice and proper

identification of the cause(s) of the defects resulting in failure of the repair and further deterioration of the original defects.

- Failure to exercise proper control procedures during the use and maintenance of the building structure.

5.28. Misuse

- Inappropriate use or change of use without proper assessment of the existing structure, resulting in overloading.
- Alterations or additions to the building or modifications to the structure without proper assessment of the existing structure, resulting in overloading or weakening of the structure.
- Abuse resulting in excessive wear, abrasion and impact.

5.29. Climatic effects

As the effects of change in temperature on movement have already been discussed (Section 5.10) only the effects of cold will be dealt with here.

It is a mistake to suppose that the generally mild climate of the British Isles is more benign to reinforced concrete structures than other harsher regimes. In fact the opposite is sometimes the case as there may be several freeze and thaw cycles during a typical British winter whereas the temperature may well stay below freezing for long periods in colder climates. More damage can be caused during the change from one state to the other than from a prolonged period in the frozen state. Water trapped in pores in the concrete expands during freezing and the forces generated can be highly disruptive. The more permeable the concrete, the greater the problem. Surface defects which can collect water can also initiate frost damage.

5.30. Ground movement

The effects of ground movement, including subsidence and frost heave, will be dealt with fully in the companion volume to this, dealing with foundations.

5.31. Water and dampness

Water and dampness cause problems in themselves only in so far as the structure is unable to resist their infiltration. Once present, however, they generally make other defects more serious.

5.32. Fire

Fire, unless the temperatures reached are above 200 °C and persist for several hours, has no great effect on the strength of concrete or on reinforcing bars having normal cover. Most damage is likely to be found on the undersides of horizontal slabs, and the colour of the concrete and extent of distortion of the reinforcement will give some indication, to a trained observer, of the temperature reached during the fire. Philip H. Perkins provides a useful guide to the investigation of fire damaged structures in *Repair, Protection and Waterproofing of Concrete Structures* (10 in bibliography).

Consideration does however need to be given to smoke deposits which may be acidic.

Select bibliography

Note The bibliography provided for Chapter 3 is also relevant to Chapter 5.

1. Berkeley K. C. G and Pathmanaban S. *Cathodic protection of reinforcement steel in concrete*. Butterworths, London, 1990.
2. British Standards Institution. *Code of practice for dead and imposed loads*, BS 6399, Part 1. BSI, London, 1984.
3. Building Research Establishment. *Shrinkage of natural aggregates in concrete*, BRE, Garston, 1968, BRE Digest 35.
4. Building Research Establishment. *Condensation*, Garston, 1972, BRE Digest 110.
5. Building Research Establishment. *Structural appraisal of buildings with long-span roofs*. BRE, Garston, 1984, BRE Digest 282.
6. Building Research Establishment. *Carbonation of concrete and its effect on durability*, BRE, Garston, 1995, BRE Digest 405.
7. Concrete Society. *The ninth international conference on alkali-aggregate reaction in concrete*, Concrete Society, London, 1992, Technical Report 104.

8. Institution of Structural Engineers. *Guidlines for the appraisal of structural components in high alumina cement concrete*. The Institution, London, 1975.
9. Institution of Structural Engineers. *Structural effects of alkali–silica reaction: technical guidance on appraisal of existing structures*. The Institution, London, 1992.
10. Perkins P. H. *Repair, protection and waterproofing of concrete structures*. Elsevier, London, 1986.
11. Williams C. Structural vibration. *Structural Survey*, 1990, **8**, No. 3, pp 280–291, and No. 4, pp 416–426.

6

Assessing and increasing strength

6.1. Introduction

The design of a new structure entails making assumptions to take account of a number of uncertainties regarding the properties of the materials to be used and the loads to be carried.

Appraisal is quite different. Here the objective is to provide a realistic picture of the adequacy of a structure which already exists, concerning which the uncertainties that had to be accommodated during design are replaced by measurable properties. The task involves not only analytical skills but also engineering judgement based on a realistic appreciation of the condition of the structure and an understanding of the manner in which it is responding to the loads which are actually carried. On the other hand the task of physically measuring all the relevant properties of the structure is likely to be extremely costly so that, again, some uncertainty — albeit on quite a different scale from those met in design — has to be catered for. Moreover the strength of concrete improves with age so that in many cases reinforced concrete structures which have been properly constructed and maintained have considerable reserves of strength. It follows that a conservative approach may be neither reasonable nor feasible but, with the aid of statistical techniques, it should be possible to present a realistic assessment of the integrity of the structure.

Because of the complex nature of structural appraisal and the serious implications of any recommendations (even negative ones) which may be made, the engineer should be aware of the legal aspects of his decisions and ensure that every stage of the assessment of strength is carried out in a thoroughly professional and responsible manner. The Introductory Guide in this series (13 in bibliography) gives guidance on legal matters.

6.2. *Load testing*

Although an expensive procedure, there will be occasions (such as when original drawings are no longer available and the complexity of reinforcement makes identifying it uncertain) when load testing is necessary. Detailed guidance is given in the Introductory Guide to this series and in J. H. Bungay's *Testing of concrete structures* (9 in bibliography). Also of specific interest will be two BRE publications: Digest 402 and IP2/95 (6 and 7 in bibliography).

6.3. *Codes of practice*

The current codes of practice represent a generally accepted level of safety provided against the uncertainties in the design and construction process. As there is presently no code specific to the assessment of building structures, it is customary to appraise existing structures by establishing whether they would comply with the level of safety in current design codes. This practice can give rise to problems because current design codes for reinforced concrete structures are more onerous than previous ones in many respects, for example in regard to shear capacity. If, however, a structure has performed adequately and without showing signs of distress — providing the engineer is confident that it has been loaded to its full design load — there can be no justification for modifying it solely on the grounds that it fails to comply with current codes of practice. The engineer may however consider warning the owner of the building that although there is no apparent danger in continuing to use the structure the situation should be re-assessed at regular intervals of, say, five years.

Before deciding to apply present day codes to old structures it should be remembered that provisions in codes of practice, both implicitly and explicitly, are based upon assumptions which are only applicable to the design methods, materials and construction practices of their day. It is therefore particularly important when applying the recommendations, formulae, tables and charts of present day codes to question critically the relevance of the underlying assumptions and, where necessary, to fall back on first principles.

The use of out-dated codes of practice also presents problems as they will not take account of technical discoveries made since they were published but they will give valuable information regarding practices at the time of design (e.g. early codes permitted the use of calcium chloride) and the properties of the materials then in use — which may differ significantly from those bearing a similar name at the time of appraisal.

The combined use of codes of practice of the present day and those in use earlier should be avoided however since there are likely to be conflicts between the two documents in underlying philosophy, assumptions and safety factors and the way in which these are made use of in calculations.

6.4. *Factors of safety*

Because, in an existing structure, the engineer is better able to establish the strength of the materials, tolerances and loading than would be the case at the design stage of a new structure it is reasonable to reduce the factors of safety below those set in codes of practice for design. Excellent guidance on this aspect is given by the Institution of Structural Engineers in *The appraisal of existing structures* (14 in bibliography).

6.5. *Improving strength or robustness*

Remedial works to deal with inadequate strength or robustness fall into the following five categories

- partial or total replacement
- duplication of individual structural elements
- strengthening of individual structural elements
- prestressing
- stiffening the structure.

Partial or total replacement

Although replacement is often the simplest and neatest solution to a lack of strength problem it is seldom the most economical, nor is it without long-term implications as any replaced member, if made of reinforced concrete, will change in size due to creep and shrinkage at a rate quite different to that of the structure within which it is placed. This is particularly significant if large compression members are to be replaced. Because such movements take place over long periods of time it is almost impossible to anticipate them in a way which will avoid some cracking or settlement. An alternative may be to use structural steelwork for the replacement member as its deformation is easily calculable and takes place almost instantly upon loads being applied.

Replacement does not always imply removing the original member as a steel framework could be built around an existing column whose new function would then be to provide stiffness to the framework.

Duplication of individual structural elements

Duplication presents similar problems to replacement and, if precautions are not taken, the situation could be reached where a new compression element creeps to such an extent that all the load would eventually be carried by the original. The problem can again be overcome if structural steelwork is used and there is the additional advantage that less space will be required than for a reinforced concrete solution.

It is however important to ensure, by careful measurement and monitoring of site operations, that the load is shared between the new and existing in the proportion intended by the designer.

Strengthening of individual structural elements

Strengthening measures can be classified as follows.

- *Providing additional reinforcement* — this is dealt with in Chapter 9.
- *Providing additional concrete section* — when providing additional concrete section care needs to be taken to ensure that the new concrete will act in conjunction with the existing concrete to

which it is providing assistance. Reinforcement at adequate intervals may be provided to tie the two sections together and the bond surface between the two concretes must be clean and free of loose material. The advice given in Chapter 8 for patch repairs should be followed when deciding on the type of concrete, admixtures and water content.

- *Improving the slenderness ratio* — the stiffness of a member can be improved either by adding concrete section, as described above, or by providing intermediate bracing.

Prestressing

It is sometimes possible to increase the load carrying capacity of beams or slabs using prestressing techniques. Tendons are anchored to the member to be strengthened and then post-tensioned. These are placed so as to subject the member to a bending moment in the reverse direction to that induced by the normal loading, which causes a reduction in overall bending moment. Needless to say, great care is necessary both in designing and choosing anchor points and in applying the correct amount of prestress. Guidance is given in an FIP publication (12 in bibliography).

Stiffening the structure

There are many ways by which a structure can be made more rigid, almost all of which can look unsightly unless great care is taken during the detailing process. Cross bracing, using structural steelwork, or infilling between columns with brickwork or blockwork are fairly obvious methods. Knee bracing or K-bracing provide less obtrusive alternatives, however, and openings may be introduced in wall panels to lessen their visual impact. In some situations diagonal prestressing cables or threaded tension rods may offer appropriate solutions.

6.6. *Foundation deficiencies and ground movement*

A companion guide in this series, which is currently in preparation, will deal with foundations and ground movements.

6.7. *Impact damage*

Much of the damage caused by impact can be classified, for purposes of treatment, under the headings already given above. There is however the special case of the buckling of a member which it would be difficult or very costly to replace. Determining its residual strength is far from easy but the following procedure may be helpful.

- Decide whether, by using jacks and temporary falsework, there is any chance of forcing the member back into its original line without causing further significant damage; in the rare case where this is possible and affordable read the advice given below as applying to the re-aligned member.
- Make a thorough examination of crack patterns, reinforcement damage or displacement and conditions of attachment at the ends.
- Consider how damage to the reinforcement and the concrete can best be repaired.
- If the member is to remain in its deformed shape calculate the loading and the stresses this would create in the repaired member in accordance with the guidelines laid down in the *Appraisal of existing structures* (14 in bibliography); appropriate eccentricities and additional bending moments must be taken into account.
- If the member is unable to carry the load safely consider the advice set out above for strengthening it; bear in mind that some form of lateral buttress or strut may be needed.

Select bibliography

1. Beckmann P. and Happold E. Appraisal — a critical process of inspection and calculation. *In. Assoc. Bridge and Structural Engineering Symp. Proc.*, Venice, 1983, **45**, 31–38.
2. British Standards Institution. *Guide to assessment of concrete strength in existing structures*, BS 6089. BSI, London, 1981.
3. British Standards Institution. *Code of practice for dead and imposed loads, BS 6399, Part 1. BSI, London, 1984.*

4. Building Research Establishment. *Loads on roofs from snow drifting against obstructions and in valleys.* BRE, Garston 1988, BRE Digest 332.
5. Building Research Establishment. *The assessment of wind loads.* BRE, Garston, 1989, BRE Digest 346.
6. Building Research Establishment. *Static load testing: concrete floor and roof structures within buildings.* BRE, Garston, 1995, BRE Digest 402.
7. Building Research Establishment. *Guide for engineers co-ordinating static load testing on building structures*, BRE, Garston, 1995, BRE Information Paper IP2/95.
8. Building Research Establishment. *Masonry and concrete structures: measuring in-situ stress and elasticity using flat jacks.* BRE, Garston, 1995, BRE Digest 409.
9. Bungay J. H. *Testing of concrete in structures.* Blackie, London, 1989, 2nd edn.
10. Comite Euro-International du Beton. *Diagnosis and Assessment of Concrete Structures — State of the Art Report.* CEB, Lausanne, 1989, Bulletin D'Information No 1921.
11. Currie R. J. Towards More Realistic Structural Evaluation. *The Structural Engineer*, 1990, **68**, No. 12, June.
12. Fédération Internationale de Précontrainte (FIP). *Repair and strengthening of concrete.* Thomas Telford, London, 1991.
13. Holland R., *et al. Appraisal and repair of building structures — introductory guide.* Thomas Telford, London, 1992.
14. Institution of Structural Engineers. *The appraisal of existing structures.* The Institution, London, 1996.
15. Institution of Structural Engineers. *Stability of buildings.* The Institution, London, 1988.
16. Tassios T. P. Physical and mathematical models for re-design of dangerous structures. *Int. Assoc. Bridge and Structural Engineering Symp. Proc.*, Venice, 1983, **45**, 259–266.

7

Treatment of cracks

7.1. To fill or not to fill?

The temptation to fill every crack in a reinforced concrete structure, for whatever reason, must be resisted. Although cracks may be deemed to be unsightly filling them may not, in every case, improve matters. Table 7.1 indicates when filling would be appropriate and when not.

Generally, it will be safe to fill those cracks where the cause of the cracking is no longer active. Cracks subject to seasonal or diurnal movement are a special case however, and are dealt with in Section 7.3 below. It should also be borne in mind that the British Standard for the design of reinforced concrete structures, BS 8110 (2 in bibliography), is drawn up on the basis that small cracks are not harmful. Except in the most aggressive situations cracks of up to 0·3 mm width, which in any case are very difficult to fill, are unlikely to lead to further deterioration. If however the risk of corrosion is to be minimized it is essential that moisture is excluded from the reinforcement so that, except in exclusively internal situations (other than where there is significant risk of condensation), protection of cracks in some way is almost always necessary. Some form of cover strip or cladding, to keep out the weather or for aesthetic reasons, might however provide an acceptable alternative.

Table 7.1 Conditions when it is either safe or unsafe to fill cracks

Safe to fill cracks (see Section 7.2 below)	*Unsafe to fill cracks*
Plastic settlement	Thermal movement (but see Section 7.3 below)
Plastic shrinkage	Corrosion related (unless cause of corrosion has been eliminated) — see Chapter 8
Early thermal contraction	Alkali-silica reaction — see Chapter 8
Drying shrinkage	Sulphates attack — see Chapter 8
Surface crazing (unless associated with alkali aggregate reaction) — but see also Chapter 8	High alumina cement (unless stresses reduced to safe level) — see Chapter 8
Delamination (unless associated with *unsafe phenomena opposite*) — but see also Chapter 8	Strength related (unless associated with appropriate strengthening) — see Chapter 8
Accidental damage	
Leakage (unless associated with *unsafe* phenomena opposite) — see Chapter 14	

7.2. *Materials for crack repair*

There are a range of materials available for the repair of cracks and a good discussion of their properties and uses is given in *The repair of concrete structures* (1 in bibliography). Many varied formulations are available from different manufacturers to cover a wide variety of climatic and other circumstances (e.g. wet surfaces) and the reader is advised to study the latest commercial literature available before making a final choice. The engineer should also be aware that it is not only the active constituents of a formulation which are important as there is often a necessarily large proportion of filler present as well. The following list, which is in descending order of cost, describes in broad terms the four categories of material most commonly employed for crack repair.

- Epoxy resin grouts — these are essentially for application by specialists but are extremely successful when used to inject narrow cracks. Very high compressive, tensile and flexural strength is gained within 48 hours and permanent flexibility, although dependent upon the formulation, can be greater than with other crack repair materials. The high strength characteristics decrease rapidly at temperatures exceeding 40 °C but the range can be extended to 80 °C with some formulations. Placing is usually by pressure or gravity injection and heat is produced during the curing process to the extent that formulations designed for use in cool weather are quite unsuitable for use during hot spells. Especial care needs to be taken when handling epoxy resins to avoid health risks and, in particular, damage to human skin.

- Polyester resin grouts and mortars — although the proportioning and mixing of the components is less critical than with epoxy resins there are other factors which call for skill in their use. Their main advantage lies in rapid gain of strength, equal to that of epoxy resins, in only a few hours. Behaviour at high temperatures is similar to that of epoxy resins but flexibility is, however, considerably less. Polyester resins produce a great deal of heat during the curing process but the greatest limitation on their use is that they also shrink during this process. They can be applied using pressure injection techniques or as a mortar. Because of their speed of setting they have been extensively used in the repair of aircraft pavements, in the installation of runway lighting and in grouting up cable ducts in pavements. Although the health risks are less than with epoxies, special care should also be taken when handling them.
- Acrylic resin grouts — very fine cracks in dry concrete may be repaired using low viscosity aqueous acrylic resin which produces a water resistant rubbery sealer that can accommodate some movement. In all but the finest of cracks several applications may be necessary.
- Polymer modified cementitious grouts and mortars — the use of polymer modifiers, such as styrene butadiene rubber (SBR), in cementitious grouts and mortars results in a material which closely resembles the concrete being repaired, in respect of strength and long term movement, but having improved characteristics with regard to bond, flexibility and time within which strength is gained.
- Cementitious grouts and mortars — these have their uses for bulk filling of cracks but, because of the shrinkage which takes place during curing, may pull away from the sides of a crack.

The engineer must satisfy himself that the durability of the proposed repair will be at least as good (unless otherwise agreed with the client) as the un-damaged part of the structure. Fortunately all the crack repair materials described above have been the subject of papers (e.g. 4 in bibliography) delivered at international symposia from which it can be deduced that, providing always that they are appropriately and properly applied, they will give good protection to the reinforcing steel.

7.3. *Cracks not associated with movement*

Injection of low-viscosity epoxy resin under pressure is the most effective method of filling cracks of between 0.1 mm and 1 mm width, particularly where it is desirable to recover the greater part of the original strength of the member. Several different resin formulations are available which should suit the needs of almost any situation, allowing injection at different temperatures, adhesion to damp or dry concrete or with a modified modulus of elasticity. The method is fully described in Phillip H. Perkin's *Repair, protection and waterproofing of reinforced concrete* (12 in bibliography). It is not possible to specify resin injection without reference to manufacturers' literature — and possibly also discussions with their representative — and the work itself should only be entrusted to firms of known competence and experience.

Before injection can commence all cracks will have to be sealed temporarily with a polymer based putty through which injection points will be provided. A number of injection points will be required for a long crack or where there is inter-connecting cracking. It is good practice, on all but the smallest of applications, to use an injection pump in which the two parts of the epoxy formulation do not come together until they reach the injection head. Vacuum impregnation may be used to advantage when there is a network of cracks to be filled. With this technique, which can only be carried out by a specialist contractor, a vacuum is created within the cracks which draws in the resin.

Cracks in the top surface of slabs may be filled with low viscosity resin flowing under the force of gravity alone. As the depth of penetration reached in these circumstances is somewhat problematic, the technique should not be used where the objectives include regaining lost structural strength.

Superfluid micro-concretes, it is claimed, offer an alternative to resin formulations for narrow cracks.

Thixotropic epoxy resins are more useful for cracks of between 1 mm and 5 mm wide in order to prevent excessive resin loss.

Cracks wider than 3 mm may be filled with cement grout but, unless some special formulation is used, the grout is likely to shrink away from the edges of the crack. Although this can subsequently be made good using epoxy resin injection it is better and quicker to use an appropriate modifier (see Section 7.2 above) in the grout or mortar initially.

For minor cracks in the top face of slabs the brushing in of cement dust followed by light spraying with water has been used. Whilst this may be effective in sealing the crack it may not significantly improve the protection of reinforcement as the depth to which the cement paste may have penetrated cannot be determined.

7.4. *Movement cracks*

If it is essential that cracks where subsequent movement (e.g. in response to temperature change) is likely should be filled then some form of filling material which will be flexible enough to accommodate the expected movement should be chosen. Such materials include natural and artificial latexes and rubber bitumens. As these are of relatively high viscosity it may be necessary to widen the crack, as shown in Fig. 7.1, before applying the filler. This has the added advantage of providing enough width for the expected movement to be accommodated without rupturing the filler.

Manufacturers of epoxy resins and polyurethanes have modified some of their products for use in moving cracks, and low viscosity versions of these may prove especially suitable for cracks of less than 0.5 mm, although the engineer needs to be satisfied that the predicted movement can be accommodated without rupture of the filler or new cracks being induced because the resin is stronger than the surrounding concrete. Whatever the material selected, it is important that the engineer responsible for the repair quantifies the expected movement and agrees with the manufacturer that the chosen material is appropriate. Manufacturers' instructions, particularly with

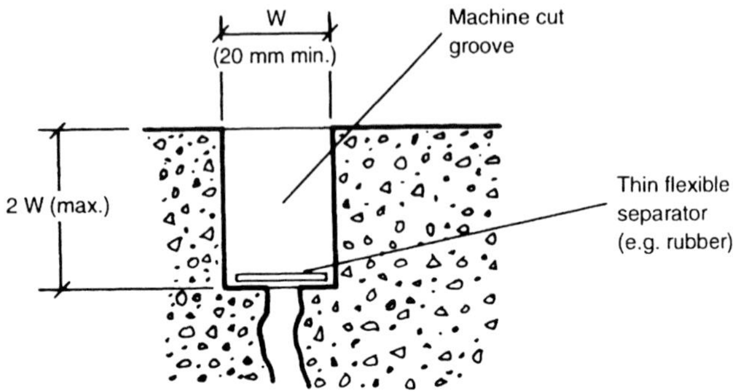

Fig. 7.1. Preparation for a repair of a crack subject to movement

respect to preparation of surfaces should moreover be carefully observed.

Select bibliography

1. Allen R. T. L. *et al. The repair of concrete structures.* Blackie, London, 1993, 2nd edn.
2. British Standards Institution. *Structural use of concrete,* BS 8110, Part 1 – *Code of Practice for design and construction.* BSI, London, 1985.
3. Building Research Establishment. *Repair and maintenance of reinforced concrete.* BRE, Garston, 1994, BRE Report 254.
4. Calder A. J. J. Corrosion protection afforded by various crack injection techniques (eds C. L. Page *et al.*). *Corrosion of reinforcement in concrete.* Elsevier Applied Science, London, 1990.
5. Concrete Repair Association. *Method of measurement for concrete repair.* The Association, Aldershot, 1994.
6. Concrete Society. *Non structural cracks in concrete.* The Society, London, 1982, Technical Report 22.
7. Concrete Society. *Grouting specifications* – extracted from the magazine 'Concrete'. The Society, London, 1993, July/Aug.
8. Construction Industry Research and Information Asssociation. *A guide to the safe use of chemicals in construction.* CIRIA, London, 1981, Publication SP16.
9. Fédération Internationale de Précontrainte (FIP). *Repair and strengthening of concrete.* Thomas Telford, London, 1991.
10. Mailvaganam N. P. *Repair and protection of concrete structures.* CRC Press, Boca Ranton FL, 1992.

11. Mays G. (ed.). *Durability of concrete structures — investigation, repair, protection.* Spon, London, 1992.
12. Perkins P. H. *Repair, protection and waterproofing of concrete structures.* Elsevier, London, 1986.
13. Puller-Strecker P. *Corrosion damaged concrete — assessment and repair.* Butterworth, London, 1987.
14. Tabor L. J. *Effective use of epoxy and polyester resins in civil engineering.* CIRIA, London, 1985.
15. Woolman R. and Hutchinson A. *Resealing buildings — a guide to good practice.* Butterworth Heineman, Oxford, 1994.

8

Other repairs to defective concrete

8.1. Thinking ahead

Before proceeding with repair the engineer must be satisfied that the durability of the chosen method will be at least as good (unless otherwise agreed with the client) as the un-damaged part of the structure. Fortunately, all the patch repair materials described below have been the subject of papers (for example, the paper by Keer, Chadwick and Thompson (14 in bibliography)) delivered at international symposia, from which it can be deduced that, providing they are appropriately and properly applied to an adequately prepared substrate, they will give good protection to reinforcing steel.

Although one could argue that it goes without saying that the cause of a defect must be established before repair work is put in hand, the number of failures of repairs directly attributable to the use of inappropriate techniques is worrying. One reason for this may be that, having gained experience with one of the many repair techniques available, an engineer may come to regard it as the panacea for all concrete problems.

Other questions which need to be resolved before the choice of technique is made are

- is the repair to be temporary or permanent?
- if it will be seen, will some form of coating be necessary for aesthetic reasons? (see Chapter 10)

- is the repair required to improve durability?
- is it intended that the repair should restore lost structural strength? (see Chapter 6).

8.2. *Replacement of concrete*

Areas from which damaged concrete has been removed may be made good by using polymer modified mortar for shallow (patch) repairs, polymer modified concrete for deep repairs and sprayed concrete for repairs extending over large areas. Where deep sections have to be replaced it may be necessary to use modified formwork techniques (see below).

Polymer modified mortar and concrete

The polymer generally used in the UK to modify mortar and concrete is styrenebutadiene rubber (SBR) which acts both as a plasticiser (thus reducing the water–cement ratio) and a bonding agent. It has the added advantage of providing a mortar or concrete with elasticity and alkalinity close to that of ordinary Portland cement concrete. In mainland Europe acrylic resin latex is the preferred polymer.

Patches exceeding 40 mm in depth are best dealt with by using a concrete mix, the size of aggregate being chosen to suit the depth of repair. A mortar mix will be found more satisfactory for shallower patches.

Although manufacturers' literature will give guidance on the use of modifying polymers, one of the essentials worth repeating here is the importance both of adequate surface preparation and of proper curing.

Other materials are marketed as being suitable modifying additives but the engineer must be satisfied that the modified product will behave in the same way as concrete with regard to movement under stress or temperature change, and that the patch will not break away from the parent concrete should impact occur. The properties of the modified material as a protection for the reinforcement also need to be considered.

Resin mortars

Epoxy resin mortars and grouts, although relatively expensive, gain strength rapidly and can be formulated to meet a variety of conditions including the following.

- Inadequate cover — epoxy resin mortars can provide adequate protection to steel reinforcement using thicknesses of only 12 mm.
- Voids — epoxy resin grouts can be used to fill voids especially where structural strength is the overriding requirement.

Polyester mortars are particularly suitable for small repairs such as chipped corners as they attain high strength rapidly. As they have higher shrinkage characteristics than epoxy resin mortars they are unsuitable for large area repairs.

Sprayed concrete

Variously known as guniting or shotcrete, the technique of spraying concrete onto a wire mesh fastened to the prepared surface of a structure is an effective method of replacing damaged concrete over large areas. There are two processes: the dry process and the wet process. In the former, which is used with aggregates smaller than 10 mm, the necessary water is added to the cement–aggregate mixture at the spray nozzle in the form of fine spray. In the latter, which can be used for coarser mixes, the water is mixed with the cement and aggregate before the concrete is pumped to the nozzle. Full details of sprayed concrete techniques are given in a chapter by W. B. Long in *The repair of concrete structures* (1 in bibliography).

Using modified formwork

Conventional formwork is designed to allow concrete to be placed economically and in a manner which will lead to proper compaction. In the repair situation it is rarely possible to make use of conventional formwork owing to access problems and the relatively small volumes of concrete to be handled. Forms may have to be designed individually for each situation and regard paid to access both for placing

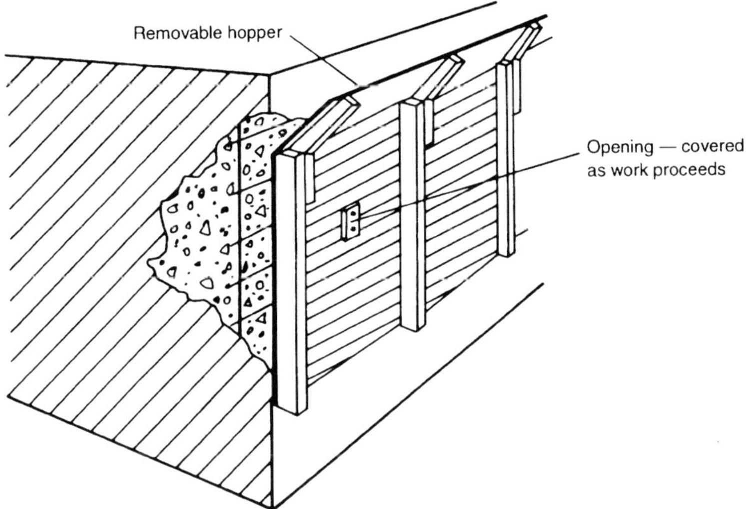

Fig. 8.1. Formwork modified to ease repair

and compaction and to the avoidance of situations which could lead to pockets of air being trapped (see Fig. 8.1). Temporary openings in the formwork, which must be sealed as the work proceeds, may be worth considering both for placement and inspection. Placing the concrete through a tube which is gradually withdrawn may help prevent separation.

The need for the repair formwork to be both rigid and well supported cannot be over-emphasized. Any movement or deflection could pull the setting concrete away from the substrate.

In designing the mix it is essential to minimize the amount by which the patch concrete will shrink or slump. Clearly, water content should be as low as practically possible and consideration should be given to the use of superplasticizers, water reducing admixtures or air entrainment.

Concrete to be replaced on the underside of a slab can effectively be delivered from above through holes cut in the slab for the purpose, if advantage is taken of superplasticizers or air-entrainment.

Grouted aggregate construction

When the vertical face of a wall, column or beam has to be cut away and a considerable thickness of concrete replaced it may be possible to erect conventional formwork, fill this with coarse aggregate and then complete the concreting by grouting up the voids. The technique is described more fully in *The repair of concrete structures* (1 in bibliography).

8.3. Preparation

Removal of defective concrete

Whenever reinforcement has corroded, the concrete should be removed from around each bar to a point at least 50 mm beyond that reached by corrosion. If the removal of pieces of concrete will affect the strength of the member being repaired it may be necessary — particularly where repairs are to be carried out in the compression zone or where bond lengths may be seriously reduced — to provide temporary support. A further matter to consider is the effect that vibrations set up during breaking out of the concrete might have on the stability of adjacent parts of the building. Slender internal masonry walls have been found to be particularly susceptible to strong repetitive vibrations and their collapse could cause serious injury.

Preparation of concrete substrate

Although some techniques are less demanding in terms of substrate preparation than others (e.g. some materials work best against a straight saw cut edge whilst others will accept a feathered edge to the repair), all require the removal of loose concrete and any impurities such as oil. An additional precaution is to provide a mechanical key to hold the patch in place by slightly undercutting the edges of the area to which the patch is to be applied.

Bonding of cementitious repairs is assisted if the surface is wetted but the practice of brushing mortar (or some other bonding agent) onto the surface in advance of the patch, although recommended by some authorities, is strongly disputed by others. The objection to bonding layers may have arisen because polymer latex bonding coats can dry out rapidly. Modified bonding agents based on spray dried copolymer powders are a more suitable alternative. Manufacturers' literature should be carefully followed however.

Preparation of reinforcement

All corrosion products must be removed by grit blasting, using a small quantity of water in the jetstream to prevent dusting. The resulting sludge can be removed by washing. Wire brushing is not effective as this does no more than polish the rust! Within four hours of cleaning, the reinforcement should be protected with a brush-on material which is compatible with the concrete repair which is to follow.

8.4. Mixing and placing of repair mortars

Manufacturers' instructions must be followed with regard to mixing of mortars if a successful repair is to be carried out. Some repair materials may entrain air if incorrectly mixed and specially designed mixers are available which reduce this risk.

Repair mortars should be built up in layers of not more than 20 mm thickness using as stiff a mix as possible and each layer should follow as soon as the preceding one is strong enough to support it.

8.5. Curing

Curing times and methods vary according to the material chosen and manufacturers' recommendations should be followed. For cementitious repairs, curing is particularly important as thin layers of mortar are liable to dry out long before the chemical reactions taking place in the patch are complete. Shading from the sun together with a temporary absorbent cover material which is kept damp should protect the work between the placing of layers of mortar, and the same procedure, or a sprayed-on curing membrane, should be used for the finished work.

8.6. Damage due to corrosion of reinforcement

Although the methods described above will be useful in replacing concrete damaged by reinforcement corrosion, additional precautions will be necessary to ensure that the repair is long lasting and the nature of these will depend upon the cause of the corrosion being recognized and dealt with. The types of corrosion and their causes are given in Chapter 5 as

- corrosion cell corrosion
- differential-aeration cell corrosion
- galvanic cell corrosion
- stray current corrosion
- bacterial corrosion.

Corrosion cell corrosion

This is the most frequently occurring type of corrosion affecting steel reinforcement in concrete but will only occur if *all* of the following three criteria are met

  - presence of water (or at least dampness)
  - presence of oxygen
  - either destruction of the passivating layer or chemical imbalance.

- Dampness — In many situations (e.g. foundations) it is rarely practicable to reduce dampness to the extent which would eliminate corrosion completely. In others, such as elements necessarily exposed to the weather, the exclusion of damp may be too expensive or aesthetically unacceptable. If coatings are used to exclude water they must be ones which do not trap moisture already held within the concrete (see Chapter 10). Above ground level inside buildings however, weather proofing and good ventilation (e.g. within ceiling voids) may significantly reduce dampness.
- Oxygen — There are no practicable methods (other than total immersion in water) by which oxygen can be excluded from concrete.
- Destruction of the Passivating Layer — this is discussed in Section 8.7 below.

- Chemical Imbalance — the most commonly occurring chemical imbalance is due to chloride attack and this is discussed in Section 8.8 below. Other chemicals can cause corrosion and the only method of elimination is to find and deal with their source and replace any concrete which has become contaminated.
- Differential-aeration cell corrosion — unless the environment in which the concrete is found can be changed, cathodic protection (see Chapter 11) offers the best means of creating the right conditions for permanence of repair.
- Galvanic cell corrosion — the metals concerned should be separated either by removing one of them or by the insertion of suitable insulation between them.
- Stray current corrosion — the services of an electrical engineer having experience of this type of problem should be sought.
- Bacterial corrosion — in the rare case where repair might be possible, the bacterial source should be removed or the concrete should be isolated from it by non-biodegradable sheeting.

8.7. Carbonation of concrete cover to reinforcement

The choice of treatment will depend upon the extent of carbonation. If limited to relatively small areas, treatment should involve cutting out affected concrete, dealing with the corroded reinforcement and making good, using one of the methods described above. The carbonated concrete can easily be identified as work progresses by means of phenolphthalein tests (see Chapter 4). The process of carbonation cannot be reversed but it can be slowed down by creating an environment within the concrete which does not encourage its spread by the use of a coating on the surface of the concrete which will restrict the amount of carbon dioxide entering the section (see Chapter 10).

Where carbonation is extensive the use of cathodic protection (see Chapter 11) may prove cost effective as this will require the removal of only the concrete cover to any corroded reinforcement, even though much of what remains may be carbonated.

8.8. Chloride contamination of concrete

Patch repair of concrete contaminated with chlorides is pointless unless the chloride levels can be reduced to acceptable levels. One method of achieving this is to use electrochemical desalination. This involves soaking the concrete in an electrolyte and attaching an anodic mesh to one surface of the concrete. An electrical current is then passed through the reinforcement (the cathode) and completed via the concrete cover and the anodic mesh, which causes chloride ions to be drawn from within the concrete. It is claimed by specialist operators that, in many cases, it takes only a few weeks to reduce chloride content to an acceptable level.

Alternatively cathodic protection may be used but this has the disadvantage of being required permanently whereas the effects of a limited period of electrochemical desalination should be permanent.

8.9. Alkali–silica reaction

To date most alkali–silica reaction problems have occurred in civil engineering structures rather than in buildings, chiefly because the chemical reaction involved requires damp conditions. If the concrete cannot be kept dry there is no known method of halting the reaction and replacement becomes the only option (see also Chapter 5).

8.10. Sulphates attack

As sulphates almost always attack foundations and are generally contained in groundwater it is almost impossible to isolate concrete from the source of the problem. Usually the only solution is to replace the concrete using a mix which is more resistant to sulphates.

8.11. Efflorescence

Efflorescence which is spread fairly evenly over the surface is often a temporary phenomenon and can usually be brushed or washed off. On the other hand, concentrations of efflorescence in a particular area may be an indication that some other defect (such as cracking or an area of porous concrete) is providing the water and chemicals required to produce the effect.

8.12. *Frost damage*

The most important factor to bear in mind when dealing with frost damage is that it could recur and cause any patch repairs to be shed. It is necessary therefore to establish the extent to which the structure could be liable to further damage by frost before proceeding with repairs. For a large area, rendering would probably prove to be the best solution whereas cutting out the suspect concrete to a depth of 50 mm and patch repairing could suffice for small areas of locally sub-standard concrete. For particularly exposed situations the use of air-entrained mortar in the repair material should be considered. Concrete which is kept dry will not be seriously affected by frost.

8.13. *Water and dampness*

The specific case of basements is considered in Chapter 14. Dampness above ground-level in buildings is caused by one, or a combination, of the following

- rain water, often from blocked gutters or downpipes, percolating, or being driven by wind, through damaged, faulty or otherwise inadequate cladding, window and door frames, flashings, joints, cable entries, etc.
- water from an internal source such as a leaking pipe, washing machine, shower, bath or sink, or from activities taking place in the building such as washing floors using excessive water
- rising damp due to inadequate damp proof courses
- condensation due to lack of ventilation or on cold surfaces
- cracks or porous areas in the concrete element.

The first measure which should be undertaken is to arrange for the cause or causes of dampness to be dealt with if at all possible. If this cannot be done the dampness should be diverted away from the affected area to somewhere less susceptible to damage. When the source of dampness can be shown to be the result of condensation, the provision of ventilation may help or, where the coldness of the concrete may be attracting the dampness, some form of insulation should be considered. It is not always easy to trace the source of water which enters a building through a flat roof as the joints or defects in the many layers which commonly make up such a roof will not necessarily coincide (see Chapter 13).

If a large area of concrete proves to be porous and the external face is accessible the situation may be improved by rendering. Otherwise, cutting out the porous concrete to a depth of 50 mm and applying patch repairs or sprayed concrete may prove satisfactory.

Once the water flowing through a crack has been reduced to no more than a 'weep' it should be possible to close the crack using techniques described earlier in this chapter. If the flow is still substantial then the advice given in Chapter 14 should be followed.

8.14. Fire damage

Any fire-damaged concrete which is friable or has a pink or pale red colouration should be suspect and cores taken to establish its residual strength. These should indicate the extent to which cutting away will be necessary before remedial works proper can begin. There are no special techniques for repairing fire damaged concrete (but see Section 9.5 below regarding reinforcement).

Select bibliography

1. Allen R. T. L. *et al. The repair of concrete structures*. Blackie, London, 1993, 2nd edn.
2. British Standards Institution. *Cleaning and surface repair of buildings*, BS 6270, Part 2: *Concrete and precast concrete masonry*. BSI, London, 1985.
3. Building Research Establishment. *Repair and maintenance of reinforced concrete*. BRE, Garston, 1994, BRE Report 254.
4. Concrete Society. *Specification for sprayed concrete*. The Society, London, 1979, CS 21.
5. Concrete Society. *Guidance notes on the measurement of sprayed concrete*. The Society, London, 1981, CS 22.

6. Concrete Society. *Repair of concrete damaged by reinforcement corrosion.* The Society, London, 1984, Technical Report 26.
7. Concrete Society. *Patch repair of reinforced concrete subject to reinforcement corrosion — Model Specification and method of measurement.* The Society, Slough, 1991, Technical Report 38.
8. Concrete Society. *Microsilica in concrete.* The Society, London, 1993, Technical Report 41.
9. Construction Industry Research and Information Association. *A guide to the safe use of chemicals in construction.* CIRIA, London, 1981, Publication SP16.
10. Concrete Repair Association. *Method of measurement for concrete repair.* The Association, Aldershot, 1994.
11. Domone P. J. L. and Jefferis S. A. (eds). *Structural Grouts.* Blackie, London, 1993.
12. Fédération Internationale de Précontrainte (FIP). *Repair and strengthening of concrete.* Thomas Telford, London, 1991.
13. Harmathy T. Z. (ed.). *Evaluation and repair of fire damage to concrete.* American Concrete Institute, Detroit, 1986.
14. Keer J. G. and Chadwick J. R. Protection of reinforcement by concrete repair materials against chloride induced corrosion. *Corrosion of reinforcement in concrete.* (eds C. L. Page *et al.*) Elsevier Applied Science, London, 1990.
15. Mays G. (ed.). *Durability of concrete structures — investigation, repair, protection.* Spon, London, 1992.
16. Perkins P. H. *Repair, protection and waterproofing of concrete structures.* Elsevier, London, 1986.
17. Perry S. H. and Holmyard J. M. *Assessment of materials for repair of damaged concrete underwater.* Department of Energy, London, 1990, Technical Report OTH 90318.
18. Puller-Strecker P. *Corrosion damaged concrete — assessment and repair.* Butterworth, London, 1987.
19. Sprayed Concrete Association. *Code of good practice.* The Association, Aldershot, 1986.
20. Tabor L. J. *Effective use of epoxy and polyester resins in civil engineering.* CIRIA, London, 1985.
21. Woolman R. and Hutchinson A. *Resealing of buildings — a guide to good practice.* Buterworth Heineman, Oxford, 1994.

9

Repair and replacement of reinforcement

9.1. Corroded reinforcement

Corrosion inevitably involves some loss of steel from the reinforcing bars. Loss of section on an individual bar can be quantified by using calipers. Alternatively a template can be made up (similar to those use by plumbers to differentiate between imperial and metric pipe sizes) from card or more durable material which, when placed against the corroded bar, indicates (by a gap between the template and the bar) the amount of steel that has been lost. Corrosion often produces pitting which will be visible if the steel is cleaned and can be allowed for when assessing the material loss.

Fortunately reinforcing bars rarely come singly and the quantity of steel required to meet the design specification is usually slightly over-provided. Moreover, although individual circumstances must be properly checked, it is often found that the initial rupturing of the concrete cover, which is generally the first observable sign of corrosion, usually occurs before significant loss of section.

If there is any doubt, the strength of the element concerned should be checked as described in Chapter 6.

All corrosion products must be removed by grit blasting, using a small quantity of water in the jetstream to prevent dusting. The resulting sludge can be removed by washing. Wire brushing is not effective as all this does is polish the rust! Within four hours of cleaning, the reinforcement should be protected with a material which is compatible with the concrete repair which is to follow.

9.2. Providing additional reinforcement

Where it is possible to place new bars alongside the existing bars and wire them into position before replacing the concrete cover this is obviously the cheapest method of repair. It may entail exposing a long length of bar in order to provide adequate bond or lap length at each end beyond the points where the extra section is required. An alternative method of providing bond is to drill holes in the concrete into which 'tails' (i.e. short bends at the end of the bar) can be inserted and grouted in. Where only a short length of bar is accessible additional bars may be welded to the existing bar providing that the heat produced by the welding process does not soften heat treated bars or damage the concrete.

An alternative involves fixing conventional reinforcement (usually in the form of a mesh) to the surface of the existing concrete section and then securing it in place by the use of sprayed-concrete techniques. This is particularly useful where large areas of slab are to be strengthened but has also been used for beams and columns.

Stainless steel should not be used in conjunction with mild steel due to the possibility of bi-metallic corrosion.

9.3. Plate bonding

A further method of providing additional tensile reinforcement is to glue mild steel plates to the tension face of the structural member using epoxy resin adhesives. A variation of this method is to secure the plates to the sides rather than the underside of a beam, to increase the shear capacity of the member.

Several manufacturers supply adhesives specifically for this purpose and their instructions for preparation and use must be followed. When choosing adhesives it is particularly important to specify the environment in which they would be expected to perform. Concern has, however, been expressed by the UK Standing Committee on Structural Safety (both in 1982 and 1989) regarding the lack of information on the long term performance of resin adhesives. Performance of epoxy adhesives in the high temperatures caused by fire are known to be unsatisfactory but there is also some evidence that long exposure to temperatures above 60 °C may lead to gradual loss of strength in some circumstances. The engineer will need to be satisfied that these objections have been met before using this technique on structurally important members in buildings.

9.4. Carbon fibre bonding

A technique is now available for bonding carbon fibre strands to the face of concrete to provide additional tensile reinforcement. As with plate bonding the engineer will need to be satisfied that manufacturers' claims as to long term durability are reasonable.

9.5. Fire damaged reinforcement

Reinforcement may be damaged by high temperature, particularly that in floor and roof slabs and thin walls. When it is suspected that the temperature of a fire has exceeded 700 °C, samples should be tested for yield strength. In some cases reinforcement may deform so badly under a combination of load

David Jones

Bonding carbon fibre plates (reproduced courtesy of New Civil Engineer)

and high temperature that the resulting sag which develops may indicate that replacement of a slab is the only practical way forward.

Select bibliography

1. Allen R. T. L. *et al. The repair of concrete structures*. Blackie, London, 1993, 2nd edn.
2. British Standards Institution. *Specification for the metal arc welding of steel for concrete reinforcement*, BS 7123. BSI, London, 1989.
3. Mays G. (ed.). *Durability of concrete structures – investigation, repair, protection*. Spon, London, 1992.
4. Perkins P. H. *Repair, protection and waterproofing of concrete structures*. Elsevier, London, 1986.
5. Puller-Strecker P. *Corrosion damaged concrete – assessment and repair*. Butterworth, London, 1987.

10

Protective coatings

10.1. Introduction

Probably because coatings are easy to apply and can be used to cover blemishes or the signs of repair activity there can be a temptation to use them without considering the consequences.

Research and experience teaches, however, that coatings should not be applied to structures containing reinforcement which is already corroding because the defect has progressed too far to be treated in so simple a way. On the other hand selected coatings on carbonated concrete where there is as yet no damage to the reinforcement (or where the damage has been properly repaired) can provide protection against the ingress of harmful gases, chemicals or water vapour.

10.2. Choice of coating material

At first sight the choice of coatings is bewildering but if the following three factors are constantly in the forefront of the engineer's mind the choice is not too difficult.

- Nature of the concrete surface.
- Purpose for which required.
- Overall cost.

Concrete surface

A porous surface will absorb greater quantities of resin which may, in turn, affect adhesion of subsequent coatings or create gaps in the cover. Specially formulated sealing or priming coats are, however, available to overcome this problem. De-bonding of the coating may occur if there is any foreign matter (e.g. oil, laitance, curing agent, form-release agent) on the surface. Only a few coatings (e.g. those which are latex based) are able to adhere to damp concrete.

Purpose of coating

There is no material manufactured that will do everything which

Before providing the protective coating those responsible for this repair should have tested the concrete for chloride content

an engineer might think desirable. It may therefore be necessary, in some cases, to specify a number of coats of different but mutually compatible materials (e.g. a primer, a barrier, a cosmetic).

Overall cost

Whether the reason for applying a coating is purely cosmetic or its primary purpose is more technical (e.g. to slow down the rate of carbonation), there will be a number of products which meet the specification. Some will have a higher material cost than others but, because of other factors, may not be the most expensive in overall cost. Cost may be affected by any or all of the following

- surface preparation
- access
- number of coats to be applied
- safety precautions required
- practical service life of coating.

The following list of the possible properties of coatings is provided to assist the engineer to judge between the merits of the many formulations offered by suppliers and manufacturers; not all are important in any one situation

- Resistance to the entry of rain water/water vapour.
- Resistance to the entry of carbon dioxide (an indication of usefulness in delaying carbonation of the concrete).
- Resistance to ultraviolet light (and therefore to chalking of any resin component of the coating).
- Resistance to the entry of sulphate or chloride ions, acid resistance.
- Permeability to water or salts escaping from the concrete (high permeability is usually sought).
- Elasticity (crack bridging capability — maximum size of crack which can be permanently covered).
- Chemical and physical compatibility with concrete.

- Adhesion to or penetration of the concrete.
- Tolerance of surface variability.
- Degree of surface preparation necessary (see below).
- Toxicity/flammability (both before and after application).
- Whether or not the substance contains solvents (an indicator as to the need for special safety precautions during applications).
- Suitable methods and ease of application (see below).
- Tolerance of typical UK weather during application — including temperature of concrete surface.
- Resistance to the growth of algae.
- Durability and length of maintenance-free life.
- Length of time before attention needed for aesthetic reasons.
- Ease of overcoating.
- Aesthetic qualities (pigment colour range, fastness of colours with time and degree of exposure, matt or gloss finishes).
- Opacity, if used to hide repair work or surface blemishes.
- Suitability for, and resistance to, removal of graffiti by organic solvents.
- Relative economy of total treatment process.
- Proven use.

If there were an ideal coating material it would be non-toxic, easy to apply, insensitive to rain and moisture during application and curing, and tolerant of severe drying conditions, dry and damp substrates, high and low temperatures, strong winds and poor surface preparation. It would last for many years without maintenance and be highly permeable in respect of water vapour passing from within the concrete but perhaps not in the reverse direction. If required to be part of a protection system for reinforcement it would be highly resistant to the passage of oxygen and carbon dioxide from the atmosphere. Of course, such a material does not exist!

10.3. Appearance

In situations where the completed work will be on view to many people appearance is of obvious importance. Glossy coatings are more likely to highlight imperfections in the concrete surface than are matt ones. Although minor blemishes can be hidden if pigmented coatings are used, significant imperfections, which cannot be ground off or filled in before coating takes place, may remain visible. The effects of any preparation which disrupts the concrete surface (e.g. grit blasting) will show through all but renderings or high-build coatings. It is advisable to trial-coat patches of about two square metres each, using several different suitable finishes or colours, before making a final decision and, if at all possible, the client or his architectural advisors should be involved in the final choice.

Consider also the possiblities of unsightly staining if, during rainfall, some areas of coating may receive water which has previously washed over surfaces visited by pigeons or surfaces which produce colourful deposits (e.g. copper). Also, some parts of a surface may be more sheltered than others so that dust is never washed away by rainwater and builds up as an unsightly mess. Choice of an appropriate colour for the coating will help in masking some of the worst effects of these phenomena.

10.4. Types of coating

Coatings may be classified in accordance with the mechanisms by which they carry out their function, as follows

- coatings/sealers — provide a thin continuous coat or seal over the concrete surface
- pore-lining treatments
- pore-blocking treatments
- renderings — provide a thick continuous coat on the concrete surface.

10.5. Coatings/sealers

These can best be described as specialist paints and are made from a number of diverse materials. Most form quite thin skins but some (e.g. chlorinated rubber) are available in high build form. Additives (e.g. fungicides) can extend the usefulness of this class of coating. There are so many formulations that it is impossible to describe each of them in detail. The list below gives a few examples and highlights some of their leading characteristics.

- Acrylics — clear or pigmented, an anti-carbonation coating, less abrasion and chemical resistant than epoxies, good inter-coat adhesion, easily re-coated, used with or without solvents.
- Acrylic/latex emulsion — pigmented, an anti-carbonation coating with good colour stability and resistance to ultraviolet light with sufficient flexibility to bridge small gaps.
- Chlorinated rubber — pigmented, anti-carbonation coating, suitable for application in cold conditions, easily cleaned, can be applied in quite thick coats.
- Epoxy polyurethane — pigmented, high-build anti-carbonation coating, flexible enough to bridge small gaps.
- Epoxy resin (two pack type) — clear or pigmented, abrasion/chemical resistant and useful for tank linings, used with or without solvents.
- Polyester resin — clear or pigmented, chemical resistant.
- Polyurethane resin (two pack type) — clear or pigmented, a useful anti-carbonation coating which also gives some protection against dilute acids (as in polluted environments) and solvents, used with or without solvents.
- Vinyl — clear or pigmented, chemical but not solvent resistant, used with or without solvents.

10.6. Pore liners

Pore lining treatments make use of hydrophobic materials which line the surfaces of pores in the concrete and repel moisture. They are marketed as

- silicones — clear
- siloxanes — clear, with or without solvents
- silanes — clear, with or without solvents.

Silanes are approved by the UK Department of Transport for use on reinforced highway structures.

10.7. Pore blockers

Pore blocking treatments react with the concrete to form compounds which to some extent fill the surface pores of the concrete. They are useful in preventing dusting of floors and increase abrasion resistance but have been used as coatings following repairs to reinforced concrete affected by corrosion. They are marketed as

- silicates — clear, with or without solvents
- silicoflourides — clear, with or without solvents
- crystal growth materials — in cementitious paste.

10.8. Renderings

Renderings are thick cementitious coatings which can be improved by the addition of polymers. There is a British Standard Code of Practice (2 in bibliography) governing their use and Building Research Establishment Digest 410 (3 in bibliography) gives further guidance.

10.9. Surface preparation

The choice of surface preparation will depend upon both the coating chosen and the condition of the concrete surface. Manufacturers can advise on the quality of surface appropriate to their products. The following list shows how specified qualities can be achieved but a great deal of time and money can be saved if the engineer notes that not all coating systems need every one of the qualities described.

- Dry — obviously, in the UK climate one cannot wait for a long period of dry weather so that the achievement of adequate dryness out of doors usually requires the provision of shelter; methods of dealing with dampness from other sources is dealt with in Chapter 8.

- Free from loose material — can be achieved by scraping and hand wire brushing; scabblers and needle guns are too powerful for this purpose.
- Free from dust — can be achieved by hand wire brushing and then brushing clean with a soft brush.
- Free from grease — this requires use of an emulsified degreaser followed by washing with clean water, although for more persistent grease or large patches of oil contamination the methods recommended by S. C. Edwards in *The repair of concrete structures* (1 in bibliography) may be necessary.
- Free from salts — if the salts are on the surface (e.g. vehicle spray) they may be removed by washing in clean water; if a little deeper, needle guns or scabblers can be used to remove the contaminated concrete which will then have to be replaced; if found throughout the concrete Chapter 8 should be consulted.
- Free from superficial laitance — will require the use of power wire brushing and, almost certainly, vacuum clearance of the dust as it is produced.
- Free from laitance and with a sharp even finish — will require the use of grit blasting on vertical and overhead surfaces (and vacuum clearance of dust), or acid etching on floors (and safe disposal of resulting slurry); high pressure water jetting (even where sand is introduced into the jetstream) is not generally suitable for this purpose.

10.10. Application of coatings

The method of application will depend largely upon the viscosity of the coating material and the area to be covered. Manufacturers' recommendations, access problems and the experience of the contractor in using various application methods should be considered. Another factor is the weather as even moderate winds can make spraying hazardous, while many formulations will not perform satisfactorily in humid conditions or if humid conditions or condensation occur within a few hours of placing.

The principal methods by which coatings may be applied are briefly described below.

- Roller application — very much the preferred method for solvent-free coatings with the advantage of speed; several coats are however likely to be needed to build up the specified thickness.
- Brush application — useful to deal with corners and other areas which roller application cannot effectively cover.
- Airless spray application — useful for large areas with good access and can be used for coating materials of medium viscosity, but may require special precautions to prevent operators and others from inhaling fumes.
- Air assisted spray application — a more accurate but slower method of delivering a controlled thickness of coating than airless spraying and one which, if pre-heating is used, can handle a wider range of paints in colder conditions; protection of operators and others even more necessary.
- Trowel application — for renderings and thixotropic high solvent-free coatings.

Select bibliography

1. Allen R. T. L., *et al. The repair of concrete structures.* Blackie, London, 1993, 2nd edn.
2. British Standards Institution. *Code of practice for external rendered finishes.* BS 5262. BSI, London, 1991.
3. Building Research Establishment. *Cementitious renders for external walls.* BRE, Garston, 1995, BRE Digest 410.
4. Institution of Civil Engineers. *Coatings for concrete and cathodic protection.* ICE, London, 1989, Report of a mission sponsored by the ICE and DTI.
5. Mailvaganam N. P. *Repair and protection of concrete structures.* CRC Press, Boca Raton FL, 1992.
6. Mays G. (ed.). *Durability of concrete structures — investigation, repair, protection.* Spon, London, 1992.
7. Vassie P. R. Concrete coatings: do they reduce ongoing corrosion of reinforcing steel.(eds C. L. Page *et al.*) *Corrosion of reinforcement in concrete.* Elsevier Applied Science, London, 1990.
8. Woodman R. and Hutchinson A. *Resealing of buildings — a guide to good practice.* Butterworth Heineman, Oxford, 1994.

11

Cathodic protection

11.1. Basic principles

Once it was realized that, in the vast majority of cases, corrosion of steel reinforcement in concrete was brought about by electrical activity, attempts were made to control differences in electrical potential and so halt corrosion. Two methods of 'cathodic' protection were thought possible — by sacrificial anode and by impressed current.

As stated already (Chapter 5) the five situations, subject to appropriate conditions, in which corrosion of reinforcement is likely to occur are as follows.

- When the chemical balance of the concrete matrix surrounding the reinforcement varies along its length (concentration cell corrosion).
- When the reinforcement is exposed to different concentrations of dissolved oxygen along its length (differential aeration cell corrosion).
- When two dissimilar metals are in contact (galvanic cell corrosion).
- When stray electrical current passes through or close to the structure.
- Where certain bacteria are active.

In such situations current flows from a positively charged anode (the corroding metal) to a cathode (metal elsewhere which is negatively charged). Anodic protection involves providing an alternative anode to the corroding reinforcement which itself will corrode, but in doing so it will protect the reinforcement — hence the term sacrificial anode. Impressed current cathodic protection, on the other hand, seeks to swamp the reinforcement system with an electrical charge (the impressed current) strong enough to cancel out all corrosion activity.

11.2. Anodic protection

The rather long official designation is 'Sacrificial-anode cathodic protection'.

One method of classifying metals is in order of 'nobility', i.e. the extent to which they will become anodic (thus sacrificial) to other metals. The less noble a metal is the greater the number of other metals it has the potential to protect. Metals which have the potential to protect mild steel (in ascending order of nobility) are

- sodium
- magnesium
- zinc — hence the use of zinc in paint, galvanizing or as a protective spray for steel
- aluminium — also used to protect steel.

Problems were encountered at one time with rapid corrosion of the steel hulls of ships in the vicinity of the bronze screw (propeller). This was recognized as galvanic corrosion with the anode and cathode being the hull and the screw respectively. Anodic protection was provided in the form of pieces of zinc secured to the hull. The zinc, being a less noble metal than mild steel, replaced the latter as the anode and the problem was solved.

Sacrificial anodes are only of limited application in the protection of reinforced concrete however, and only where at least part of the concrete is submerged in water or buried in waterlogged soil. They have, for example, been used to prevent corrosion in water storage tanks and marine structures.

11.3. Protection by impressed current

Impressed current cathodic protection has been successfully applied to reinforced concrete structures in the following situations.

- Protection of reinforced concrete bridge decks affected by road salt.
- Protection of reinforced concrete building structures where the concrete contained calcium chloride.
- Protection of reinforced concrete building structures where there was significant carbonation of the concrete cover.
- Protection of marine structures.

Within the compass of this book it is not possible to give more than a brief outline of the principles involved in impressed current protection and to provide the interested reader with a list of more specialized publications.

Impressed current protection involves the provision of a network of (anodic) conductors on the surface of the concrete and the passing of a very low voltage direct electrical current through the concrete between the network and the reinforcement. The network is connected to the positive terminal of the supply and the reinforcement to the negative terminal. The reinforcement must first be tested for electrical continuity. If this is lacking it may be provided by welding individual bars together or linking discontinuous reinforcement together with cables. Except where precast concrete has been used this is rarely a major problem.

Control of the system is very important as too low a current would not inhibit corrosion sufficiently and too high a current may cause some deterioration of the concrete or even hydrogen embrittlement of the steel. As uniform a current as possible is desirable as large differences of potential along a reinforcement system could, in itself, cause corrosion. To achieve this it may be necessary to provide a higher voltage where dense reinforcement occurs. Control is usually exercised via reference electrodes installed at intervals and linked to a data logging device. This device could, in turn, be linked to the office of the engineer by telephone so that the frequency of monitoring visits can be reduced. Notwithstanding this visits should be made to assess the effectiveness of the system at monthly intervals over the first year and twice a year thereafter.

All the required remedial works, such as replacement of defective concrete, should be carried out, and any unsound concrete removed, before the cathodic protection system is installed. It should, however, be borne in mind that epoxy resins are electrically non-conductive and may, particularly if injected into cracks parallel with the surface, affect the success of the protection system.

Although design and installation of the system should always be carried out by specialists the engineer needs to consider which of the alternatives available would be most convenient and cost effective for the individual application. The following methods of providing the anodic half of the system have been used.

- Paint — containing carbon particles, and thus capable of conducting electricity, is supplied with power through wires at approximately 1.2 m centres.
- Flame-sprayed zinc — with power again provided through wires.
- Mesh — (somewhat like chicken wire) with a metal oxide surface is mechanically fastened to the concrete surface.
- Cables — of flexible polymeric strand or copper, mechanically fastened to the face of the concrete in a serpentine configuration.
- Ceramic plates — spaced at intervals and connected by cables.

Before activating the system it should be checked for short circuits such as might be caused by individual bars or tying wire reaching close to the concrete surface or via embedded items such as holding down bolts, drains and joint materials. Also the tightness of all connections should be checked.

All the systems described require protection in the form of a spray-on electrical coating, as described in Chapter 10, in accordance with the manufacturers' recommendations.

Select bibliography

1. British Standards Institution. *Cathodic protection*, BS 7361, *Code of Practice for land and marine applications* Part 1. BSI, London, 1991.
2. Concrete Society. *Cathodic protection of reinforced concrete: Report of a joint Concrete Society and Corrosion Engineering Association working party*. Concrete Society, London, 1989. Technical Report 36.
3. Institution of Civil Engineers. *Coatings for concrete and cathodic protection*. ICE, London, 1989, Report of a mission sponsored by the ICE and DTI.
4. Mays G. (ed.). *Durability of concrete structures — investigation, repair, protection*. Spon, London, 1992.
5. Perkins P.H. *Repair, protection and waterproofing of concrete structures*. Elsevier, London, 1986.

12

Prestressed concrete

12.1. Introduction

Although much of the advice given elsewhere in this guide is directly applicable to prestressed concrete there are a number of additional precautions which must be taken.

Unless the affected face is unseen, corrosion damage in normal reinforced concrete is likely to be visible long before it has seriously affected the integrity of a structure. Moreover, progressive loss of strength is likely to be detectable through evidence such as cracks or deflection long before complete structural failure. Conversely, although cracks and deflections are also useful indicators of problems in prestressed concrete the period of time between their appearance and structural failure may be comparatively short for the following reasons.

- The high tensile stress levels in prestressing steel, even in the unloaded condition.
- In pre-tensioned slabs, the high bond stresses developed between the steel and the surrounding concrete.

Moreover corrosion in prestressing steel is often difficult to detect as the under-side of floor and roof slabs, where most use is made of pre-tensioned products, are often masked by ceilings and, almost without exception in post-tensioned work, the tendons are contained in grouted or un-grouted ducts where it is almost impossible for corrosion to produce signs which will be detected at the surface.

12.2. Corrosion in prestressed concrete

As has been already stated (Chapter 5) certain conditions are necessary before corrosion of reinforcement will take place. For most types of corrosion these are moisture, oxygen and destruction of the protective passivating layer (or of any alternative protection provided). With certain grades of steel used in post-tensioned work however, a type of corrosion (hydrogen induced stress corrosion — see following paragraph) which does not require the presence of oxygen could also be a threat. On the other hand all types of corrosion require the presence of moisture and the first objective of any investigation should be to establish whether it is safe to assume that the tendons are in a permanently dry state.

Hydrogen-induced stress corrosion

A pre-requisite for hydrogen-induced stress corrosion is some other form of corrosion mechanism which is capable of producing hydrogen as a by-product. In addition to the more frequently encountered mechanisms of corrosion the possiblity of stray electric currents or a badly designed or improperly installed cathodic protection system should always be considered. The free hydrogen is re-absorbed into the metal and diffused along grain boundaries where it creates planes of weakness — a sort of unzipping process. Eventually, if unchecked, it will lead to brittle fracture. The process is more fully described in a paper by B. Isecke (8 in bibliography). Because a knowledge of the types of steel which might be affected is required, an engineer encountering corrosion of tendons should seek metallurgical advice.

12.3. Pre-tensioned slabs and beams

By far the most numerous examples of prestressed concrete construction in building structures in the UK consist of pretensioned products in the form of

- thin precast planks in floors and flat roofs whose strength is increased by the use of in-situ structural concrete topping
- shallow precast beams, solid or hollow, placed side by side with in-situ non-structural topping providing a continuous upper surface and filling the small gaps between the beams
- shallow precast solid or hollow beams laid in parallel, with gaps between which are spanned by lines of hollow clay-ware, or lightweight, concrete pots; over the whole a structural in-situ topping is placed which has the effect of converting the beams into structural tee-sections.
- deep precast beams at relatively wide centres, the gaps between being spanned by precast planks (see above) and the whole covered with an in-situ structural concrete topping.

Pre-tensioned products generally make use of wire strand which is tensioned and anchored externally before the concrete is placed. Once the concrete has set the strand is severed from its anchorages so that the tension is then held by bond with the concrete. Corrosion will of course destroy this bond and, if allowed to progress significantly, will lead eventually to loss of prestress. It follows that any appraisal of a deteriorating pre-tensioned member must include an assessment of the possible loss of strength caused in this way.

One of the problems in dealing with loss of bond as a result of the corrosion of prestressing steel is that the preparation for repair is likely to cause further de-bonding. The engineer must take a hard decision, where there is considerable de-bonding, as to whether or not repair can be regarded as a safe option given the possibility of injury to operators should the tensile forces in a tendon suddenly overcome the remaining bond strength.

12.4. Post-tensioned construction

Although the appraising engineer must always be alive to the possibility that post-tensioned construction has been used in the building which he is investigating this is unlikely to be the case in buildings designed in the UK except in the following situations

- where comparatively long distances (i.e. in terms of those normally met in building structures) are spanned
- where large moments or eccentricities on compression members make its use economically viable
- where architectural design requirements make conventional reinforced concrete unsuitable.

The great problem for the appraising engineer is in deciding what degree of deterioration, if any, has occurred to prestressing tendons within the ducts which contain them. This, in turn, depends on two factors

- the original form of protection provided
- whether any of the causes of corrosion, delineated above and described more fully in Chapter 5, are present.

Cable protection systems

Two principal methods of cable protection have been used — grouted and un-grouted. The former, which is by far the most common, relies on the properties of the grouted material (normally cementitious) to protect the cable in a similar way to the protection afforded to reinforcement by its cover. The effectiveness of the grout depends mainly therefore (assuming that it was of suitable composition) on whether it has penetrated completely.

Detection of corrosion

Detection of corrosion is by no means easy. In the case of un-grouted tendons it may be possible to explore the cable duct using optical probes. With grouted tendons, on the other hand, it is best to first explore those areas where the grout would have had the most difficulty in penetrating, as it is in such locations that the tendons are most likely to be unprotected. One of the more common locations for corrosion of tendons is in the vicinity of anchorages. Techniques for locating voids, such as ultrasonic pulse velocity testing described in Chapter 4, could be used although their effectiveness may be limited by the material of the duct walls. X-ray techniques offer a good chance of locating voids in ducts but the equipment is bulky and the operation hazardous as well as expensive.

Selected cutting out may offer the most effective method of locating voids in grouted ducts although, in some situations, it may be aesthetically unacceptable. Drilling and the insertion of optical probes provides an alternative. If the exact location of the ducts is unknown, holes should be drilled in stages and the resulting powder examined for traces of the duct material before the next stage is commenced. Damage to the tendons must be avoided at all costs.

Inspection of the tendons within any voids found must be thorough as any small corrosion pit could be masking more severe stress corrosion as discussed earlier. Load testing should not be attempted on post-tensioned structures without the oversight of an engineer who has a thorough practical experience of the behaviour of such structures, as there is a very real danger that sudden and dramatic collapse could be initiated. Falsework, strong enough to hold up the structure should it collapse during testing, should be considered.

Repair

Repair will, of course, depend upon the type and seriousness of any defects found and, insofar as the defects are in the concrete or require voids to be filled, should be based on guidance given elsewhere in this guide. Repair of tendons should never be attempted without metallurgical advice. Replacement or duplication in almost all cases offers safer alternatives. Chapters 6 and 9 give guidance on the strengthening of structures.

Anchorages

Because of the high stresses induced in the concrete around anchorages some micro-cracking is almost inevitable. This is usually harmless but should be monitored. Any tendency for the cracks to widen should receive immediate consideration. As already mentioned, one of the more common locations for corrosion of tendons is in the vicinity of anchorages. Unless exposure is severe, corrosion of the anchorages themselves (in buildings) is unlikely to become a serious problem.

Select bibliography

1. Concrete Society. *Durability of tendons in prestressed concrete*. Concrete Society, London, 1982, Technical Report 20.
2. Fédération Internationale de Précontrainte (FIP), *Corrosion protection of unbonded tendons*. Thomas Telford, London, 1986.
3. Fédération Internationale de Précontrainte (FIP), *Demolition of reinforced and prestressed concrete structures*. FIP, Madrid, 1982.
4. Fédération Internationale de Précontrainte (FIP), *Precast, prestressed hollow core floors*. Thomas Telford, London, 1988.
5. Fédération Internationale de Précontrainte (FIP), *Repair and strengthening of concrete structures*. Thomas Telford, London, 1986.
6. Fédération Internationale de Précontrainte (FIP), *Stress corrosion of pre-stressing steel*. FIP, Madrid, 1981.
7. Halvorsen G. T. and Burns N. H. (eds). *Cracking in pre-stessed concrete structures*. American Concrete Institute, Detroit, 1989.
8. Isecke B. Long term behaviour of materials in a prestressed concrete bridge. *Corrosion of reinforcement in concrete*. (eds C. L. Page *et al.*) Elsevier Applied Science, London, 1990.

13

Floors and roofs

13.1. Introduction

Many of the defects in reinforced concrete floors and roofs can be rectified using methods described elsewhere in this guide. This chapter concentrates on those defects which are generally only found in floor and roof slabs. These can be listed as follows (the reader is reminded however that this guide is restricted to defects likely to be associated with building structures — and hence external roads and pavements are not dealt with).

- Ground floors
  - subsidence
- All floors
  - wear
  - delamination of surface
  - unsatisfactory finish
- Roofs
  - rainwater ingress
- General
  - defective expansion joints.

13.2. Subsidence of ground floor slabs

Where subsidence is a problem it is necessary first of all to establish whether the movement has stopped or is still continuing. This may be done by taking selective levels at intervals of time. In many cases, however, the amount of use of a ground floor slab may be such that there will not be sufficient time to undertake monitoring. In such cases investigation of the ground supporting the slab can be carried out by drilling bore holes and examining the subsequent cores. These should show whether the causes are localized (e.g. lack of consolidation or the action of water) or deep seated (e.g. over-loading of the supporting strata or the results of mining activity).

If the cause is deep-seated reference should be made to a companion volume to this dealing with the *Appraisal and Repair of Foundations* which is now in preparation. Localized causes of subsidence can however be dealt with in a number of ways, the choice between which will depend upon economic rather than technical considerations.

Reconstruction of slab and replacement of unsatisfactory supporting ground

Although this may be an expensive solution it is likely to be the most successful method in many cases.

Pressure grouting under the slab

This has been used in many cases where the floor slab could not be made available for long periods of time. Care must however be taken to protect drains and ducts from also becoming filled with grout. Unless the top surface of the slab can be raised beforehand to the intended finished level this method may leave awkward areas of the surface to be made good (see Section 13.3 below).

Vacuum grouting under the slab

This is a very effective development of pressure grouting whereby a vacuum is created under the slab by suction which draws the grout in to fill any voids.

Expansive grouts

Some manufacturers offer additives which cause grout to expand during the setting process. These can be effective in ensuring that voids are filled if used with care and in accordance with the manufacturers' instructions.

13.3. Surface wear and tear of concrete floors

Several methods are available for dealing with the surface wear and tear of concrete floor surfaces and the choice of an appropriate method will depend upon the following factors

- time available for repair
- extent of damage
- intended use of repaired surface
- general regularity or unevenness of floor surface.

Methods of repair are as follows.

- Local repairs — localized defects can be patched after using a concrete saw set to the depth of the defect to cut out a square at least 75 mm larger than the defect in each direction. The area within the square should be reduced to a common level and cleaned of all loose material. Because of the difficulty of batching small quantities of concrete/mortar on site, it is preferable to use a prepared specialist mix containing a polymer such as SBR (styrene butadiene rubber) latex for the repair.
- Screeds — screeds are thin (e.g. 38 mm) layers of sand/cement concrete laid on top of an existing concrete slab to provide a more regular finish. Before placing the screed the surface of the slab requires preparation to remove dirt, oil and other contaminants and to provide an adequate key. Concrete planing machines or portable shot blasting plants can be used to achieve this. The use of an appropriate bonding agent between the prepared slab and the screed is essential.
- Toppings — the term 'toppings' covers a variety of layers applied to the surface of concrete floors to improve their

wearing qualities. The following will prove useful in the repair situation.

- Granolithic screeds — these are hard wearing screeds whose aggregate consists of granite, flint, quartzite or other hard wearing minerals.
- Unbonded toppings — where the existing concrete has to be increased in thickness, shrinkage cracks can be avoided if the necessary topping, providing that it is at least 100 mm in thickness, is regarded as a separate slab. This can be achieved by separating it from the existing slab by a layer of polythene on a damp proof membrane. A further, more hard wearing, topping of 10 to 15 mm in thickness can be placed and compacted into its surface.
- Rapid curing toppings — a number of proprietary toppings are available, based usually on resins or polymers, which cure much more rapidly than cementitious toppings. They are particularly useful where time to carry out the work is very limited. These are well described in *The repair of concrete structures* (1 in bibliography).

13.4. Delamination of floor surfaces

Screeds may become detached from the concrete slab which supports them due to poor bonding or thermal or moisture shrinkage. They may also fail because of poor compaction or inadequate mixing causing variable cement content throughout the screed.

Unless delamination is extremely localized, patch repairs are unlikely to be successful. On the other hand the un-damaged portion of a bonded screed or topping may not break away easily from the supporting slab. It is important therefore to carry out a thorough examination before deciding upon a final specification for repair work.

If the concrete is dry the injection of a low viscosity aqueous resin, specially formulated for the purpose, can provide a useful method of re-bonding a screed.

13.5. *Unsatisfactory finishes to concrete floors*

The surface of a floor may be unsatisfactory for one or more of the following reasons

- it is uneven
- it is not level
- it is not sufficiently resistant to abrasion

Uneven floors

When a floor is level but uneven it is possible to overcome the defect by planing the floor to remove the worst high spots and then using one of the methods outlined above to provide a screed or topping.

Unlevel floors

Unlevel floors often present more of a difficulty to the repairer than uneven floors, particularly if the difference in level from one edge to the other is considerable. If the slab is satisfactory in all other respects, and there are no objections to raising the finished level, the use of an unbonded topping may be appropriate. On the other hand, if the difference in level across the slab is small, limited planing of high spots and the provision of a bonded screed may suffice. In extreme cases reconstruction of the slab may be called for.

Improving abrasion resistance

The ability of a floor to withstand abrasion can be increased by polymer impregnation. The method is described in *The Plant Engineer's Handbook* (10 in bibliography) in a chapter on industrial flooring by J. D. N. Shaw.

13.6. *Rainwater ingress through roof slabs*

When investigating suspected rainwater ingress through a roof slab it may save much time if three other possible sources of dampness can be eliminated first.

- condensation — thorough drying out and ventilation of the space below the roof slab before and during any 'rain' test would limit the possibility of condensation occurring during the test.
- interstitial condensation — i.e. condensation occurring between the layers which make up the roof. Broadly speaking, if there is thermal insulation above the concrete slab interstitial condensation is unlikely to be a problem because the concrete will be within a few degrees of room temperature. On the other hand, concrete above the thermal insulation may well be subject to interstitial condensation if there is no adequate moisture barrier below it. A common situation where interstitial condensation might occur is where an insulated ceiling is hung below a reinforced concrete roof slab and the void so created is inadequately ventilated.
- faulty rainwater goods — gutters choked with leaves or other debris, cracked or blocked downpipes or broken supports to either may cause dampness.

Having reached this point the great difficulty then is in deciding where the water enters the roof structure as this may be some distance from the signs of damp so far discovered. First of all the more usual points of entry should be examined for defects and, in dry weather, each should be subject to a 'rain' test (i.e. using a hose or watering can) in turn, to see if this can replicate the effects already noticed. The more usual points of entry for rain water are

- vertical surfaces — parapets, chimney stacks, flanking walls — both flashings and damp proof courses should be examined.
- joints of all kinds — but especially movement joints
- service entries — ducts, chimneys, pipes and (especially) rainwater downpipes
- areas subject to concentrated rainwater discharges — i.e. where pipes discharge (or gutters overflow) onto a roof.

If this fails to indicate the source of leakage, or if leakage continues after repair of the defects already located, the next step is to carry out a thorough examination of the roof covering for signs of damage or deterioration. It is worth mentioning here that ponding on a roof rarely indicates leakage whereas, on the other hand, the lack of ponding in a low spot may indeed be the source of the problem. Any defects then found should be repaired using the methods appropriate to the materials involved or by re-covering the roof if the defects are extensive. An excellent book dealing in more detail with this subject is K. F. Endean's *Investigating rainwater penetration of modern buildings* (8 in bibliography).

An interesting expert system for diagnosing the cause of defects in flat roofs has been developed by Dr R. J. Allwood at the University of Technology, Loughborough. At the time of writing it is understood that attempts are being made to launch the system commercially.

13.7. Defective expansion joints

Typically, expansion joints consist of a gap containing a compressive filler material (e.g. fibre board, cellular rubber, granulated cork board) and finished with a sealer (typically a hot applied thermoplastic material or a mastic substance but sometimes a preformed strip secured by adhesives) in a wider groove at the surface. Expansion joints may also incorporate a water bar either within the thickness of the construction or at the surface.

Leakage of water through expansion joints is a frequently occurring problem and may be caused by any of the following defects as illustrated in Fig. 13.1

- rupture of the sealant
- detachment of the sealer from the concrete
- puncture of the water-bar
- defects in the concrete alongside the joint
- a combination of any or all of these defects.

Rupture of the sealant

The reason for the damage needs to be ascertained before an appropriate remedial strategy is formulated. Although the cause could be the result of physical damage or chemical spillage (e.g.

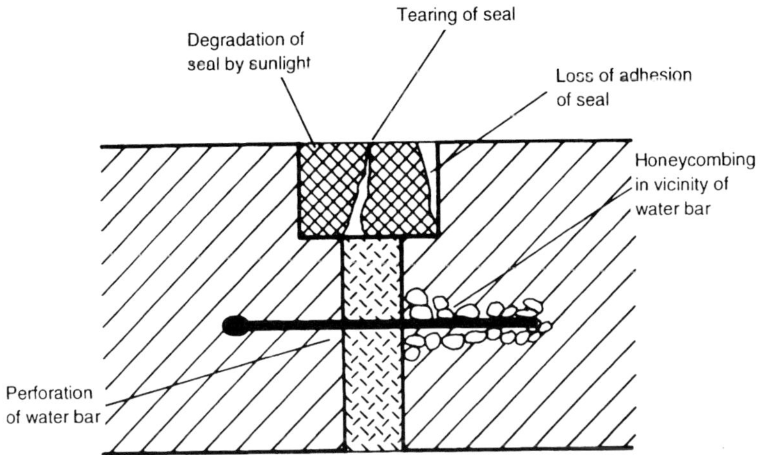

Fig. 13.1. Typical causes of leakage through expansion joints

fuel oils such as petrol and kerosene can dissolve bitumen compounds) or tearing as a result of excessive movement of the joint, more common causes are brittleness brought about by gradual degradation through long exposure to sunlight or the sealant never having been properly installed.

To repair any damage the filler must first of all be inspected to ensure that hard materials, which could prevent the gap from closing, have not been picked up as a result of the damage. The faces to which the new sealant is to adhere must be cleaned (in accordance with the instructions of the manufacturer of the sealant chosen) and the sealant applied, again in accordance with instructions.

Although it goes without saying that a most important consideration in the formulation of sealing materials is that they should be capable of expanding and contracting without rupture, it is not always appreciated that if there is not sufficient clear distance between the points of adhesion of the sealer to the concrete the strains set up by movement may exceed the strain capability of the sealing material. It is important therefore to ensure that the sealant groove is wide enough and that the sealant does not adhere to the filler (the use of a thin flexible separator, as shown in Fig. 7.1, should be considered).

Replacement filler materials can be selected from the following list and more detailed information regarding each will be found in

Philip H. Perkin's *Repair, protection and waterproofing of concrete structures* (9 in bibliography).

- Hot applied thermoplastics.
- Cold applied thermoplastics.
- Thermo-setting compounds.
- Mastics.
- Pre-formed strips secured by adhesives.

Detachment of the sealer from the concrete

This defect is normally the result of poor control during construction. The defective sealer should be removed and the repair carried out as described above.

Puncture of the water-bar

Again this usually results from construction malpractice. If this defect is suspected the filler should be removed and a water-test applied to locate the defect. Examination using an endoscope should reveal its exact position. Repair will not be easy and as each situation will be different the solution is perhaps best left to the ingenuity of the engineer concerned. What must be remembered however is that whatever repair is made must not restrict movement of the joint.

Defects in the concrete alongside the joint

This is a very common cause of problems (e.g. local honeycombing) and one which should be suspected before assuming that, for example, a water bar is punctured. Repair methods described elsewhere in this guide are applicable but care must be taken to prevent solid materials from blocking movement of the joint.

Select bibliography

1. Allen R. T. L. *et al. The repair of concrete structures*. Blackie, London, 1993, 2nd edn.

2. British Standards Institution. *Screeds, bases and in-situ floorings*, BS 8204, *Code of Practice for concrete bases and screeds to receive in-situ floorings*, Part 1. BSI, London, 1987.
3. British Standards Institution. *Screeds, bases and in-situ floorings*, BS 8204, *Code of Practice for concrete wearing surfaces*, Part 2. BSI, London, 1987.
4. British Standards Institution. *Screeds, bases and in-situ floorings*, BS 8204, *Code of Practice for polymer modified cementitious working surfaces*, Part 3. BSI, London, 1993.
5. British Standards Institution. *Screeds, bases and in-situ floorings*, BS 8204, *Code of Practice for terrazo working surfaces*, Part 4. BSI, London, 1993.
6. Concrete Society. *Concrete industrial ground floors*. The Society, London, 1988. Report 34.
7. Domone P. J. L. and Jefferis S. A. (eds). *Structural Grouts*. Blackie, London, 1993.
8. Endean K. F. *Investigating rainwater penetration of modern buildings*. Gower, Aldershot, 1995.
9. Perkins P. H. *Repair, protection and waterproofing of concrete structures*. Elsevier, London, 1986.
10. Snow D. A. (ed.). *Plant Engineers Handbook*. Butterworth Heinemann, London, 1991.

14

Basements

14.1. Introduction

Structures below ground level present special problems with regard to appraisal and repair, the first being that it is rarely possible to gain access to the outer face of the structure. Consequently investigations and repairs can usually be carried out only from the inside. When dealing with leaks, particularly those involving large quantities of water, situations may however be encountered (for example, where poorly designed or constructed basements are founded below the water table, close to a river) when it would not be possible to prevent the ingress of water except by working on the outer face. As methods of repair described elsewhere in this guide are generally applicable to basements and, as the problems of leakage and dampness are especially severe in underground structures, this chapter will concentrate on the latter topics.

14.2. Classification of basements

After many years of dealing with problem basements in public buildings the PSA (the former agency responsible for the provision and maintenance of the UK Government civil and defence estate) issued guidance to clients who were contemplating acquiring or having property built which would include underground facilities. This guidance took account of both the

Table 14.1 Categories of basement

Use to which basement is to be put	Location		
	Well-drained site on sand/gravel permanently above natural water table	Site normally, above natural water table	Site permanently below natural water table or close to river
Archives	C	D	E
Offices or accommodation for people	B	C	D
Garages or storage of items not affected by damp	A	B	C

use to which the basement was to be put and the location in terms of the likely threat from water. Included in the advice was the following classification of basement types (Table 14.1).

Although the classification was intended to demonstrate the escalating cost (on a sliding scale rising from A to E) of providing satisfactory accommodation, it nevertheless can serve as a convenient method of classifying basements for the consideration of repairs.

- Category A basements — any leaks or seepage are likely to be the result of damage to services in the ground near the basement or to periodic heavy rain. Leaking services should be dealt with and rainwater channelled away — externally if possible — although internal collection may present an acceptable and cheaper alternative. Dampness may be the result of condensation which can often be cured by better ventilation.

- Category B basements — leaks and seepage are unlikely to occur throughout the year (unless the result of damage to services) and can be dealt with as described later in this chapter. Water from leaks which cannot be stopped may be collected in open channels and led to a sump to be pumped away. Water entering through cracks in the roof can similarly be collected in gutters and discharged to the sump. The engineer needs to be sure that the continual passage of water through the structure will not affect its integrity during the remainder of its expected life or become a health hazard.

Good ventilation is essential and, if dampness persists, it may be necessary to line the walls and ceiling with insulating panels incorporating a damp-proof membrane. Battens used to support the panels should treated to protect them from deterioration and there should be a ventilated gap of at least 75 mm between the panels and the structure. A lining of this kind should be provided with removable panels to permit periodic inspection of the worst leaks.

If all leaks have been brought under control it may be cheaper to provide a suitable rendering rather than a lining and there are several proprietary renderings available which provide both insulation and some resistance to damp.

The floor may require a topping to resist the upward pressure of dampness or to raise the working level above channels provided around the edge to take away water. There are several suitable proprietary materials available.

- Category C basements — the advice given for Category B basements should be followed with the exception that water should be led through a closed pipe system to the sump to avoid creating unwanted condensation. The pipes may be grouted into the structure at the leaks.

  Also, in basements of this category, except for those containing only vehicles and goods not affected by damp, it is likely that dry lining and air conditioning will be required to deal with the problems caused by a damp atmosphere and the advice of an experienced building services engineer should be sought. It is also possible that a raised and ventilated floor would prove more suitable than a floor topping.

- Category D and E basements — that such basements can be successfully provided, if adequate precautions are taken in design and during construction, is evidenced by the Public Record Office which is founded in the Thames gravels at Kew and contains many basements storing invaluable archive material. That such basements can be successfully repaired is another matter altogether.

It should be borne in mind that not all the accommodation in a building needs to reflect the same high standards as the best and it may be worth considering with the client whether the areas which are likely to prove the most difficult to repair could be down-graded to less demanding uses.

Repair techniques would be identical to those recommended for Category C basements.

14.3. Sources of leaks

If at all possible the problem of leaks should be tackled at their source rather than where they show within the building. Drains and sewers should be checked for blockages and water loss checks over a period of time can be undertaken with the co-operation of the owner of the service. Invasion by tree roots is a common cause of damage to and blockage of drains and sewers. High-pressure water jetting is sometimes used to clear blockages in pipes but can result in total and explosive rupture of small drains. Tracing dies (but only those types which are not affected by the alkalinity of concrete) can be used to confirm a suspected source. In the case of mains water services the Water Companies have specialized equipment for tracing leaks.

Another cause of excessive water pressure in relatively shallow structures could be the blocking of the natural drainage of the site or of the drainage provided to carry water away from the external face of the walls of a basement. Observation of changing water levels in bore holes or trial pits could help to confirm such a hypothesis.

The roofs of subterranean structures are particularly prone to leakage and the resulting drips of water may contain salts which are potentially harmful to materials stored below.

Not all underground structures rely on reinforced concrete alone for their resistance to the passage of water. Some designs incorporate a form of protective layer on the outside face consisting of asphalt or bitumen. Once such a layer becomes ruptured it is rarely possible to repair it from inside and, moreover, any water penetrating the layer may travel some considerable distance before finding a defect in the concrete through which it can escape.

14.4. Construction imperfections

A number of construction imperfections can lead to leaks or seepage by providing easy paths for water to pass through the structure. These, and their treatment, are discussed below with reference to seepage and minor leaks but the treatment of major leaks is dealt with later in the chapter.

- Construction joints and shrinkage cracks — poor preparation or shrinkage after construction make construction joints a major source of seepage and minor leaks. They and other shrinkage cracks can usually be dealt with by cleaning out the joint with compressed air and injecting resin as described in Chapter 7 using a material chosen for its ability to adhere to wet concrete.
- Movement joints and restraint cracks — where the crack is likely to continue to move, the filler chosen must also be capable of expanding and contracting. Polyurethanes and bitumen compounds may be suitable. Where the movement is significant it may be necessary to widen the crack over about 20 mm of its depth before sealing it to provide enough thickness of fill material. To achieve good adhesion of the filler a clean regular surface is required and it is therefore best to use a saw, rather than a chisel, to widen cracks. Further advice on the treatment of defective expansion joints is given in Chapter 13.
- Voids — voids may be formed where water bars become displaced, where air pockets prevent concrete from completely filling the forms, where shutter-tie holes are not completely filled or where plastic settlement occurs under reinforcing bars. In the last case voids may be long enough to allow water to travel far from its entry into the structure to its point of discharge. Large voids should be repaired using patch repair techniques but smaller ones, particularly those along reinforcing bars, are better treated using epoxy resin injection.
- Honeycombed concrete — although it is fairly obvious that honeycombed concrete will allow water to pass through, the honeycombing itself may not be visible as it could be located either on the unseen side of the structure or in its centre. Tapping the surface with a hammer may however detect

sub-surface voids. An alternative technique is to drill a small hole and insert an optical probe, guidance regarding which will be found in the Introductory Guide to this series of books.

Honeycombed concrete on or close to the near face should be cut away and replaced with a patch repair. The surest method of dealing with honeycombing on the far face would also be cutting out and replacement as one can never be completely certain how effective other methods will be. Where cutting out would be excessively expensive, injection of epoxy or polyurethane or water-based acrylic resin formulations may be worth trying.

14.5. Leak sealing

As water will find the easiest passage through any structure it can be expected that the sealing of those leaks which are apparent at any one time may lead to leakage through other imperfections.

Although leaks are most effectively sealed from the outside (i.e. the pressure side) by the use of impermeable sheets or coatings, this will be possible only with shallow basements and where access is not restricted. This chapter therefore deals with the sealing of leaks from the inside for which the techniques described below are available. All the techniques described involve the use of experienced and skilled operators. For readers seeking a more detailed treatment of the topic, the chapter by S. C. Edwards in *The repair of concrete structures* (1 in bibliography) is recommended.

- Pressure grouting — polymer-modified cement grout is pumped through several 50 mm holes drilled in the structure at about one metre away from the leak to form a curtain of solid cementitious material between the structure and the surrounding soil. This technique has been in use for several years and can be effective. Unfortunately the side effects can be disastrous if the pressurised grout finds its way into pipes, sewers or service ducts.

- Caulking — low pressure leaks can sometimes be stopped by drilling out or sawing the crack to a regular shape and either forcing in a bitumen plug or injecting a resin of putty consistency capable of setting in water. Patch repairs using ultra-fast setting cement mortar can sometimes be used in conjunction with caulking.

- Self-sealing lances — sufficient holes of a specified diameter are drilled around the leak to allow the water to flow away whilst the original leak is repaired using methods described elsewhere in this book. Proprietary lances with water-tight seals are then secured in the pre-drilled holes and resin is injected through them at high pressure. Once this has set the injection equipment is uncoupled leaving the lances embedded in the structure. Some assessment of the likely water-pressure is required in order to choose the appropriate size of lance.
- Indirect injection — where there are several leaks in a concrete structure and its condition leads the engineer to consider that treating each of them directly is likely to lead to further leaks becoming active, certain proprietary resin formulations can be injected at intervals through the structure which may then be carried along by the flow of water and deposited in the cracks.

Select bibliography

1. Allen R. T. L. *et al. The repair of concrete structures*. Blackie, London, 1993, 2nd edn.
2. Domone P. J. L. and Jefferis S. A. (eds). *Structural grouts*. Blackie, London, 1993.
3. Perkins P. H. *Repair, protection and waterproofing of concrete structures*. Elsevier, London, 1986.

15

Periodic inspection and maintenance

15.1. Introduction

It is unfortunate that, more often than not, engineers are called upon to inspect and appraise concrete structures only when the signs of deterioration or distress are obvious and the condition has approached a serious state requiring extensive and costly repairs. The need for periodic inspection and regular maintenance of a concrete building structure during its life cannot be over-emphasized.

The risks which could be run if adequate maintenance is not provided can be listed as follows.

- Unseen deterioration (e.g. carbonation), if not stopped in time, could lead to unsightly cracking and spalling.
- Cracking or deformation, if not attended to early enough, could lead to expensive repairs becoming necessary.
- Cracked and spalling concrete could cause a hazard to users and pedestrians if pieces of concrete become detached.
- The load carrying capacity of the structure may be impaired if deterioration is allowed to proceed unchecked.
- The appearance of the structure is likely to deteriorate.
- The market value of the structure may be adversely affected.

15.2. Periodic inspections

Inspection of reinforced concrete structures should be planned as part of the regular maintenance programme for all building structures whether or not they appear to be free of defects on the surface. It is often when a planned inspection is carried out that defects which are hidden from sight, or potential defects, are detected. No period of more than five years should elapse between inspections and this period should be reduced to as little as one year when serious defects are already known to exist or in the period following repairs.

The aim of a periodic inspection is to identify, at an early stage, symptoms of deterioration and structural distress so that the situation can be prevented from becoming worse. Such an inspection would, in the main, entail a visual condition survey of the structure by an experienced engineer but, in certain circumstances, this would need to be supplemented as follows

- by hammer tapping of concrete surfaces to detect delamination of concrete cover where visual inspection indicates a problem may be developing
- by removal of loose concrete cover at selected areas to establish the stage which any reinforcement corrosion may have reached
- by measurement of selected crack widths to determine if there has been movement since the last inspection
- by chipping away plaster or other finishes to determine whether any observed cracks in them are reflective of cracks or movement in the structure proper.

15.3. Objectives of the periodic inspection

During each periodic inspection the objectives should include the following

- Condition of the structure
  - identification of the types of structural defects present or suspected
  - identification of any signs of structural distress or deformation

  - identification of any signs of material deterioration
  - identification of any damage which may have occurred as a result of impact, explosion, fire or from other causes
  - monitoring of the performance of repairs or protective measures already undertaken.
- Loading on the structure
  - identification of any change of use or any misuse which might have resulted in overloading
  - identification of any addition or alteration which might have caused overloading or other adverse effects on the structure.

The guidance given in Chapters 2 and 3 will be useful in planning and carrying out the periodic inspection.

15.4. The report on the periodic inspection

All observations should be classified and recorded in a systematic manner with photographic evidence where necessary. The report should state what further investigations, if any, are considered necessary. Where the report is to include recommendations on non-structural matters (e.g. appearance) the standard which is acceptable to the client needs to be agreed beforehand.

In its BMI Special Report 167 *Condition Surveys* (7 in bibliography) the Chartered Institute of Building lists four general levels of condition.

- As new with the expectancy that with proper maintenance the building will provide a satisfactory standard of service.
- Satisfactory, safe, with only minor deterioration which can be dealt with within existing maintenance budgets.
- The building is operational but major repair or replacement will be necessary within a reasonably short period with costs outside the current maintenance programme.
- Inoperable, unsafe, with risk of immediate breakdown requiring urgent expenditure outside the current maintenance programme.

Although the above wording may not be directly relevant to reinforced concrete a broadly similar classification of this kind is helpful in drawing busy managers' attention to the more urgent issues. Of course, if the engineer discovers that something is unsafe it should be reported immediately and action taken to protect users and the public.

Further guidance on report writing is given in the Introductory Guide to this series (10 in bibliography).

15.5. Maintenance of reinforced concrete building structures

Well-compacted dense reinforced concrete having proper cover to the reinforcement requires less maintenance than almost any other material used in the construction of buildings. This statement in no way contradicts the opening paragraph of this book as, unfortunately, not all concrete is dense or well compacted nor is the correct amount of reinforcement cover always provided. Moreover aggregates can vary from the ideal and assumptions made by designers regarding exposure and use are not always borne out in practice.

Any effective maintenance programme for reinforced concrete must be based on a system of periodic inspections as described above. Following from these the following actions will need to be considered.

- Repairing damaged or deteriorating reinforced concrete as and when recommended by the inspecting engineer.
- Strengthening the structure, or modifying the loading regime, in order to deal with any over-stressing or damage which the inspecting engineer may report.
- Replacing or repairing damaged or defective jointing, sealing, covering or coating materials.
- Cleaning out and repairing gutters, down-pipes, etc. where overflows or leaks could cause deterioration of the concrete.
- Clearing away any debris or growths which might otherwise lead to contamination or deterioration of the concrete.
- Where carbonation is found to be advancing at a significant rate consideration should be given to applying an appropriate

coating to the surface — particularly if the structure is exposed to an industrial atmosphere.

- In some situations (e.g. alkali–silicate recreation) the progress of degradation can be arrested, or at least retarded, by providing appropriate protection.

Except where significant deterioration has already taken place a system of period inspections by an experienced engineer coupled to a planned maintenance programme is likely to keep all reinforced concrete building structures in the best condition possible at minimum cost because any problems are thereby discovered before they have spread too far to be remedied without extensive cutting away. In the case of reinforced concrete regular health checks avoid major surgery!

Select bibliography

1. British Standards Institution. *Cleaning and surface repair of buildings*, BS 6270, *Concrete and pre-cast concrete masonry*, Part 2. BSI, London, 1985.
2. Building Research Establishment. *Concrete of lichens, moulds and similar growths*. BRE, Garston, 1992, BRE Digest 370.
3. Building Research Establishment. *Cleaning external surfaces of buildings*. BRE, Garston, 1983, BRE Digest 280.
4. Building Research Establishment. *Controlling mould growth by using fungicidal paint*. BRE, Garston, 1995, BRE Information Paper 12/95.
5. Building Research Establishment. *Repair and maintenance of reinforced concrete*. BRE, Garston, 1994, BRE Report 254.
6. Campbell-Allen D. and Roper H. *Concrete structures: materials, maintenance and repair*. Longman, London, 1991.
7. Chartered Institute of Building. *Condition surveys*. Chartered Institute of Building Maintenance Information Service, Ascot, 1988, BMI Special Report No. 167.
8. Concrete Society. *Design, construction and maintenance of concrete storage structures*. The Society, London, 1985, CS 006.
9. Fédération Internationale de Précontrainte (FIP). *Inspection and maintenance of reinforced and prestressed concrete*. Thomas Telford, London, 1986.
10. Holland R., *et al.* (eds). *Appraisal and repair of building structures — Introductory Guide*. Thomas Telford, London, 1992.

11. Taylor G. *Maintenance and repair of structural concrete.* Chartered Institute of Building Maintenance Information Service, Ascot, 1981.
12. Whitford M. J. *Getting rid of graffiti: a practical guide to graffiti removal and anti-graffiti protection.* Spon, London, 1991.

Index

Abrasion resistance, 128
Access, 4
Acid attack, 66
Acrylic coatings, 109
Acrylic resin grouts, 83
Acrylic resin latex, 89
Acrylic/latex emulsion coatings, 109
Aggregates, 44, 59
Aggressive materials, 66
Algae, 67
Alkali–silica reaction, 30, 34, 64, 96
Alumina, *see* High alumina cement
Aluminium, 60, 114
Ammonium compounds, 66
Anchorages, 122
Anodic protection, 114
Appraisal
  Initial, 7–19
  procedure, 4
Architectural finishes, 12, 20
Atmospheric pollution, 66–67
Autogenous healing, 23

Bacterial corrosion, 61
Basements, 134–140
Blowholes, 37
Bond strength, 32
Bonding agents, 89, 93
Brass, 61
Brick slips, 34
Brief, 4–5
Bronze, 61
Buckling, 36, 47
Bulging, 33

Cable protection systems, 121
Carbon fibre bonding, 102
Carbonated concrete, 64
Carbonation, 29, 48, 62–63, 95
  depth, 15–16, 44
Cast iron, 61
Cathodic protection, 96, 113–117
Caulking, 139
Caustic soda, 66
Cement content, 63
Cement type, 44, 50
Cementitious grouts, 83
Cementitious mortars, 83
Ceramic tile coverings, 34
Chemical analysis, 44, 50
Chloride
  contamination, 29, 44, 64, 96
  content, 15–16
Chlorides, 49, 62
Chlorinated rubber coatings, 109
Cladding, 2
Climatic effects, 71
Coatings, 104–112
  application, 111
Codes of practice, 75
Compressive strength, 32
Concentration cell corrosion, 60
Concrete cover, 15–16
Concrete strength, 16, 44, 49
Condensation, 64, 97–98, 129
Construction defects, 36
Construction deficiencies, 69
Construction joints, 34–35, 138
Copper, 61
Core samples, 17
Corrosion, 29, 59, 60, 61–64, 100

Corrosion (*continued*)
in prestressed concrete, 119
of reinforcement, 48, 93–95
Cover, *see also* Concrete cover, 37, 44, 49, 63
Cover meter survey, 44
Cover meters, 16, 49, 51
Crack injection, 84
Crack repair, 82
Cracks, 21–23, 26–32, 35, 38, 43–44, 46, 81–87, 122, 138
Creep, 32, 77
Crystal growth materials, 110
Curing, 93
Curtain walls, 2

Dampness, 35, 47, 72, 97
Defects, *see* Construction defects
Deficiencies
construction, 69
design, 68
Deflections, 36, 47
Delamination, 33, 44, 46
of floor surfaces, 127
Density, 44, 50
Depth of carbonation, 44
Design deficiencies, 68
Desk-top study, 4
Differential settlement, 32
Differential-aeration cell corrosion, 60
Dirt, 22, 67
Distortion, 36, 38, 47
Distress
Signs of, 20–39
Drying shrinkage, 28, 58

Early thermal contraction, 28, 58
Efflorescence, 34, 66, 96
Electrical potential, 63
Electrochemical desalination, 96
Endoscope, 16
Endoscopic survey, 44
Epoxy polyurethene coatings, 109
Epoxy resin adhesives, 101
Epoxy resin coatings, 109
Epoxy resin grouts, 82, 90
Epoxy resin mortars, 63, 90
Epoxy resins, 116
Ettringite, 66
European practice, 3
Excessive voidage, 50
Expansion joints, 130–132
Expansive grouts, 126
Exploratory investigations, 5
Exudation, 34

Façade, 2
Factors of safety, 76
Field specification, 9, 42
Fire, 72, 102
damage, 98
Flexural strength, 31
Floors, 124–133
Formwork ties, 35
Foundation deficiencies, 70, 78
Frost
damage, 67, 71, 97
heave, 71
Fruit juices, 66

Galvanic cell corrosion, 60
Global survey, 7
Gold, 61
Granolithic screeds, 127
Grit blasting, 93
Ground movement, 71, 78
Grounded aggregate construction, 92
Grout loss, 35
Grouts, 82, 83, 90, 125, 126
Guniting, 90
Gunmetal, 61
Gypsum plaster, 66

Half-cell potential mapping, 44, 48
Hammer tapping, 16
Hammer testing, 44
High alumina cement, 31, 67
Historic buildings, 5
Honeycombed concrete, 138
Honeycombing, 35–36
Hydrogen embrittlement, 115
Hydrogen-induced stress corrosion, 119

Impact damage, 79
Impressed current cathodic protection, 115
In-situ testing, 44
Indirect injection, 140
Initial appraisal, 7–19
Inspection, *see* Periodic inspection
Instrumentation, 5
Interstitial condensation, 129
Intrinsic cracks, 23, 26, 30

Lamination, 16
Lead, 61
Leak sealing, 139
Leakage, 35
Leaks, 137
Legal
  aspects, 5
  matters, 75
Lichen, 67
Lime bloom, 67
Load testing, 5, 44, 75

Magnesium, 114
Maintenance, 144–145
  deficiencies, 70
  programme, 70
Micro-concretes, 85
Micro-cracking, 122
Milk, 66
Misuse, 71
Mixing and placing of repair mortars, 93
Modified formwork, 90–91
Monitoring, 5
Mortars, 63, 83, 89, 90, 93
Mould, 47
Movement, 22, 29, 59
  cracks, 85
  of ground, 71, 78
  joints, 35, 138

Non-concrete components, 38
Non-destructive tests, 15–16

Overload, 32

Passivation layer, 62–64
Patch repairs, 88–90
Periodic inspection, 70, 141–145
Permanent formwork, 27
Permeability, 44, 50, 55
Petrographic analysis, 44, 50
Phenolphthalein, 16, 48
  test, 44
Phospher-bronze, 61
Plastic settlement, 27, 35, 57
Plastic shrinkage, 27, 58
Plasticiser, 89
Plate bonding, 101
Pollution
  atmospheric, 66–67
Polyester mortars, 90
Polyester resin coatings, 109
Polyester resin grouts, 83
Polyester resin mortars, 83
Polymer modified cementitious grouts, 83
Polymer modified cementitious mortars, 83
Polymer modified concrete, 89
Polymer modified mortar, 89
Polyurethene resin coatings, 109
Pop-outs, 32
Post-tensioned construction, 120–121
Powdered surfaces, 37
Precast beams, 120
Precast planks, 120
Precast structures, 2
Preparation
  of concrete substrate, 92
  of reinforcement, 93
  of surface, 110–111
Pressure grouting, 125, 139
Prestressed concrete, 118–123
Prestressing, 78
Pre-tensioned slabs and beams, 119
Protective coatings, 104–112
Pulverised fuel ash concrete, 3
Pyrites, 34

Quality of concrete, 55
Quantab test strip, 16, 44, 49

Rainwater ingress, 128
Rapid curing toppings, 127

Reinforcement, 93, 100–103
Reinforcement corrosion, 48
Renderings, 110
Repairs to defective concrete, 88–99
Replacement materials, 50
Replacement of concrete, 88–89
Report, 5
Resin mortars, 63, 90
Resistivity
  measurement, 44
  testing, 48
Restraint cracks, 138
Robustness, 76
Roofs, 124–133
Rust stains, 34

Safety, 4, 76
Schmidt hammer, 16, 44, 49–50
Scouring, 37
Screeds, 126, 127
Self-sealing lances, 140
Service life, 70
Settlement
  Differential, 32
Shear strength, 31
Shift, 22
Shotcrete, 90
Shrinkable aggregates, 59
Shrinkage, 27, 28, 58, 77
Shrinkage cracks, 138
Signs of distress, 20, 22–39
Silanes, 109–110
Silicates, 110
Silicoflourides, 110
Silicones, 109
Siloxanes, 109
Silver, 61
Smoke deposits, 72
Sodium, 114
Spalls, 33
Sprayed concrete, 90
Stainless steel, 49, 61, 101
Stalactites, 34, 67
Stereoscopic examination, 44, 50
Stray current corrosion, 61
Strength, 31, 32, 76
  assessment, 74–80
  of concrete, 49
Strengthening, 77
Structural deformation, 36
Structure
  stiffening, 78
Styrenebutadiene rubber, 89
Subsidence, 71
  of ground floor slabs, 125
Sugar, 64
Sulphates attack, 30, 66, 96
Sulphates contamination, 44
Sulphur dioxide, 67
Surface crazing, 29, 59
Surface preparation, 110–111
Suspended ceilings, 2

Tearing, 36
Tensile strength, 31
Thermal movement, 29, 59
Toppings, 126, 127
Torsional strength, 32

Ultrasonic pulse velocity, 16
Ultrasonic pulse velocity (UPV)
    testing, 44, 50
Unbonded toppings, 127
Uneven floors, 128
Unlevel floors, 128
Unusual materials or structures, 5

Vacuum
  grouting, 125
  impregnation, 84
Velocity testing, 44
Vibrations, 36, 47
Vinyl coatings, 109
Visual conditions survey, 8, 12
Voidage, 50
Voids, 35–36, 44, 138

Water, 66, 72, 97
  absorption, 44, 50
  bars, 35, 132
Water retaining structures, 35
Weathering, 33
Windsor probe, 44
Wire brushing, 93, 100

Zinc, 60, 114